The Politics of Decentralization

Forests, People and Power

Edited by

Carol J. Pierce Colfer and Doris Capistrano

Routledge
Taylor & Francis Group
LONDON AND NEW YORK

To those whose voices have not yet been heard.

First published by Earthscan in the UK and USA in 2005

For a full list of Earthscan publications please contact:
Earthscan
2 Park Square, Milton Park, Abingdon, Oxon OX14 4RN
711 Third Avenue, New York, NY, 10017, USA

Routledge is an imprint of the Taylor & Francis Group, an informa business

First issued in paperback 2016

A catalogue record for this book is available from the British Library

Library of Congress Cataloging-in-Publication Data has been applied for

Typesetting by Fish Books
Cover design by Susanne Harris

ISBN13: 978-1-84407-205-7 (hbk)
ISBN13: 978-1-138-99510-9 (pbk)

Contents

PART II COUNTRY CASES 139

PART III COMMUNITY VOICES 253

List of Contributors

Steve Amooti Nsita (steven@nfa.org.ug) is a coordinator, Natural Forest Management, National Forestry Authority, PO Box 70863 Kampala, Uganda.

Edmond Barrow is a forest officer, World Conservation Union (IUCN) Regional Office, Nairobi.

Jürgen Blaser (jblaser@intercooperation.ch) is with Intercooperation, an organization for development and cooperation, Bern, Switzerland.

Timothy C. Boyce is a state forester in Alabama.

Doris Capistrano (d.capistrano@cgiar.org) is director, Forests and Governance Programme, Center for International Forestry Research (CIFOR), Bogor, Indonesia.

Mike Carroll is director, Minnesota Department of Natural Resources, Division of Forestry, Minnesota.

Cherukat Chandrasekharan is a consultant in Kerala, India.

Carol J. Pierce Colfer is a principal scientist, Forests and Governance Programme, Center for International Forestry Research (CIFOR), Bogor, Indonesia.

Arnoldo Contreras-Hermosilla is forest trends fellow, Forest Trends, Washington, DC.

Silvel Elías (silvelias@yahoo.com) is a professor, Facultad de Agronomía, Universidad de San Carlos, Guatemala, and a doctoral candidate, University of Toulouse, France.

Christopher Elliott is director, Forest for Life Programme, World Wide Fund for Nature (WWF) International, Gland, Switzerland.

Ian Ferguson (iansf@unimelb.edu.au) is emeritus professor, University of Melbourne, Parkville, Victoria, Australia.

Hans M. Gregersen (hans@walk-about.net) is chair, Standing Panel on Impact Assessment, Science Council, Consultative Group on International Agricultural Research (CGIAR), Solvang, California.

Steve Gretzinger works for the World Wide Fund for Nature (WWF) Regional Office for Central America, San José, Costa Rica.

Mandy Haggith (hag@worldforests.org) is a researcher with Worldforests, 95 Achmelvich, Lochinver, Sutherland, Scotland IV27 4JB.

Steven Hlambela is a resident of Chizvirizvi resettlement, Chiredzi district, Zimbabwe.

Witness Kozanayi (kozanayi@yahoo.com or mandondo@africaonline.co.zw) is a consultant at the Center for International Forestry Research (CIFOR), Eastern and Southern Africa Regional Office, Harare, Zimbabwe.

Christian Küchli (Christian.kuechli@buwal.admin.ch) works with the Swiss Agency for the Environment, Forests and Landscape (BUWAL), Bern, Switzerland.

Anne M. Larson (alarson@tmx.com.ni) is a research associate, Center for International Forestry Research (CIFOR), Managua, Nicaragua.

Gary Lettman is a principal economist, Oregon Department of Forestry, Oregon.

Ted L. Lorensen is a senior staff member, Oregon Department of Forestry, Oregon.

Douglas W. MacCleery is a senior policy analyst, US Department of Agriculture (USDA) Forest Service, Washington, DC, US.

Stewart Maginnis is with the Forest Conservation Programme, World Conservation Union (IUCN), Gland, Switzerland.

Natalia V. Malysheva (nat-malysheva@yandex.ru) is head of department, All-Russian Research Institute of Silviculture and Forestry Mechanization, Ministry of Natural Resources of the Russian Federation, Moscow.

Thomas McShane works for World Wide Fund for Nature (WWF) International, Gland, Switzerland.

Felix S. Mirasol Jr. (jqmmm@philcom.ph) is protected area superintendent of Mount Kitanglad Range Natural Park, Bukidnon, the Philippines.

Merilio G. Morell (merilio.morell@fao.org) is an institutions forestry officer, Forestry and Institutions Service (FONP), Forestry Department, United Nations Food and Agriculture Organization (FAO), Rome, Italy.

Pablo Pacheco (p.pacheco@cgiar.org) is a PhD candidate, Clark University, consultant to the Center for International Forestry Research (CIFOR), Bogor, Indonesia, and research associate at the Institute of Environmental Research for Amazônia (IPAM), Belem, Brazil.

Lauren Phillips is an intern, Forest Trends, Washington, DC.

Jesse C. Ribot (ribotj@wwic.si.edu, jesser@wri.org) is a fellow at the Woodrow Wilson International Center for Scholars, Washington, DC, and a senior associate in the Institutions and Governance Program, World Resources Institute, 10 G Street, NE, Suite 800, Washington, DC 20002.

Bill Ritchie (bill@worldforests.org) is a researcher with Worldforests, 95 Achmelvich, Lochinver, Sutherland, Scotland IV27 4JB.

Gerald A. Rose (jerryrose@uplogon.com) is a forest sustainability consultant and former Director of Forest/State Forester in Minnesota.

Oppon Sasu (sasuoppon@yahoo.com) is a professional forester as well as a project manager. He holds an MSc in forest management from the University of Aberdeen and an MBA in project management from the University of Ghana. Currently he is the Business Planning Manager of the Forest Services Division of the Forestry Commission, Ghana, and Team Leader for the High Forest Resource Project.

Adolino L. Saway Alyas Datu Makapukaw is chief, Council of Elders of Mount Kitanglad Range Natural Park, Bukidnon, the Philippines.

Jeffrey Sayer (jsayer@wwfint.org) is a senior associate, World Wide Fund for Nature (WWF) International, 1196 Gland, Switzerland.

Gill Shepherd is with the World Conservation Union (IUCN) Commission on Ecosystem Management, London.

Bruce Springer is a forest management chief in Alabama.

Wandojo Siswanto (wandojo@dephut.cbn.net.id) is secretary, Agency for Forestry Planning, in the Indonesian Ministry of Forestry, Jakarta, Indonesia.

Wahjudi Wardojo is secretary general of the Indonesian Ministry of Forestry, Jakarta, Indonesia.

Andy White is a senior director, Policy and Market Analysis, Forest Trends, Washington, DC.

Hannah Wittman is a PhD candidate, Department of Development Sociology, Cornell University, Ithaca, New York, US.

David C. Zumeta is executive director, Minnesota Forest Resources Council, Minnesota.

Foreword

Forests are profoundly local. Each one is unique and is adapted to its particular climate, soils and topography, and its history, as well as its use, depend heavily upon its specific social and economic context. People with formal forest management responsibilities are more successful in their endeavours if they tailor their efforts to local conditions. It is hard to maintain forests long without local support, and to achieve such support, communities must feel they benefit.

At the same time, forests are truly global. The whole world benefits from their rich biological and cultural diversity. Changes in forest size and composition affect the global climate. Animals and plants move from one place to another without regard for national borders. The same applies to the smoke and haze from forest fires, the sedimentation of rivers and the lack of drinking water caused by deforestation.

National governments bear the responsibility for the future of their forests. Their citizens look to them to ensure that forest resources provide economic growth and jobs, and to enforce the laws and protect the environment.

The challenge is to find a governance framework that can balance the various local, national and global interests related to forests. Everyone agrees that local groups should be allowed to come up with solutions that reflect their own needs and circumstances; but regional, national and global concerns must also be addressed.

This book grew out of an initiative by the governments of Indonesia and Switzerland in the framework of the United Nations Forum on Forests to promote a global dialogue about these issues. As part of that initiative, the two governments co-sponsored a workshop in Interlaken, Switzerland, called Decentralization, Federal Systems of Forestry and National Forest Programmes. The Interlaken workshop was held on 27–30 April 2004, and brought together more than 160 participants from 51 countries, representing well over 70 per cent of the global forest surface.

The Interlaken workshop was designed to give high-level decision-makers concerned with forests and other key stakeholders the opportunity to share experiences about decentralization in the forest sector and to find out about recent research on the topic. The outcome was a lively and interesting discussion, the results of which are reflected in an official UN report.

Now, with the publication of this book, we hope to share some of the background information, experiences and conclusions with a wider audience. Given the urgent need to find the right balance among local, regional, national and global governance of forests, we thought it important to make this material available to a much wider audience.

As the co-sponsors of this initiative, we would also like to give a special word of thanks to the Center for International Forestry Research (CIFOR) and Switzerland's organization for development and cooperation, Intercooperation, which have provided much of the technical and administrative support for this initiative, to the Food and Agriculture Organization (FAO) of the United Nations, the Program on Forests (PROFOR) at the World Bank, the International Tropical Timber Organization (ITTO), the World Wide Fund for Nature (WWF), the World Resources Institute (WRI), and the governments of Canada, Japan, the UK and the US, which provided valuable technical and financial support for the initiative. Our thanks also go to Carol J. Pierce Colfer and Doris Capistrano of CIFOR for editing the volume.

If the Interlaken workshop made anything clear, it was that the search for the appropriate balance between the authorities at different levels responsible for forests is never-ending and constantly evolving. It probably would not surprise anyone that Indonesia is grappling with difficult issues after only a few years of a major process of decentralization. Yet, Switzerland, which has been refining its decentralized approach for several hundred years, is still trying to adjust the balance of powers to get things right. All of us are in a process of constantly learning. This book is part of that process; we hope it can help us move forward, for the sake of the world's forests and the people who depend upon them.

Wahjudi Wardojo, secretary general of the Ministry of Forestry of Indonesia, and Philippe Roch, director, Swiss Agency for the Environment, Forests and Landscape
March 2005

Preface

The idea for the Interlaken workshop on Decentralization, Federal Systems of Forestry and National Forest Programmes was initiated by Jagmohan Maini, then head and coordinator of the United Nations Forum on Forests (UNFF) secretariat in New York. The governments of Indonesia and Switzerland organized the workshop as a country-led initiative in support of UNFF, and the plan was officially announced in 2002 at UNFF 2 (the second formal meeting of UNFF). The workshop took place in Interlaken, Switzerland, from 27–30 April 2004, the week preceding UNFF 4.

The workshop was co-hosted by the Center for International Forestry Research (CIFOR) and co-sponsored by the governments of Brazil, Canada, Ghana, Japan, the Russian Federation, Uganda, the UK and the US. Technical or financial support was provided by the secretariat of UNFF, the secretariat of the Convention on Biological Diversity (CBD), the National Forest Programme Facility of the Food and Agriculture Organization (FAO), the World Bank's Program on Forests, the International Tropical Timber Organization (ITTO), the World Wide Fund for Nature (WWF), the World Resources Institute (WRI), the Swiss Agency for the Environment, Forests and Landscape (BUWAL), the Swiss Development Cooperation (SDC), and Intercooperation, a Swiss organization for development and cooperation, which also took responsibility for the logistics for the workshop. A field day, representing an integral programme element in the Interlaken workshop and also described in this book, was organized by BUWAL in close cooperation with the Forest Service of the canton of Berne.

About 160 people from 51 countries participated in the workshop, representing 70 per cent of the global forest area. It is expected that the initiative on decentralization in the forest sector, brought forward at Interlaken, will further influence the debate on the links among sustainable forest management, sustainable development goals, in general, and goals of highest societal importance, such as poverty alleviation.

The chapters in this book are organized to reflect the three kinds of experience shared at the workshop. The first part looks at decentralization from a thematic perspective, examining such issues as biodiversity, democracy and geography. The second part takes a national perspective examining several country cases. The final section presents three community perspectives on the experience of decentralization. In this way, we hoped to reflect the diversity that characterizes decentralization experiences around the world, and capture any generalizations that could be gleaned from this diversity.

Acknowledgements

Our sincere and heartfelt thanks go to Rahayu Koesnadi, Linda Yuliani, Luluk Suhada and Michael Hailu at CIFOR who have been particularly important in keeping things moving along smoothly. We also want to thank Olivia Vent and Sally Atwater who struggled with our various kinds of English, and finally, Camille Adamson, Rob West and Michael Fell from Earthscan, particularly for their willingness to work together to ensure a timely production process.

List of Acronyms and Abbreviations

AA	appropriate authority
AMEDIKP	Association of Eulalensis Women for Pixan – Komop Development (Guatemala)
AMUNIC	Nicaraguan Association of Municipal Goverments
BOLFOR	Project for Sustainable Forest Management (Bolivia)
BOSCOM	Communal Forests Office, National Forest Institute (Guatemala)
BUWAL	Swiss Agency for the Environment, Forests and Landscape
CAMPFIRE	Communal Areas Management Programme for Indigenous Resources (Zimbabwe)
CBD	Convention on Biological Diversity
CBO	community-based organization
CGIAR	Consultative Group on International Agricultural Research
CIFOR	Center for International Forestry Research
CIIFAD	Cornell Institute for International Food and Development
CND	National Commission for Decentralization
CONAP	National Council for Protected Areas (Guatemala)
CPF	Collaborative Partnership on Forests
CSM	Cellule Stratégie et Méthode (Burkina Faso)
CSO	civil society organization
DFID	UK Department for International Development
EU	European Union
FAO	United Nations Food and Agriculture Organization
FLASCO	Facultad Latinamericana de Ciencias Sociales (Guatemala)
FODECOM	Fonds de Dévelopment des Communes (Burkina Faso)
G8	Group of 8 (industrialized countries, including the Russian Federation)
GEF	Global Environmental Facility
GPRS	Ghana Poverty Reduction Strategy
GTZ	German Agency for Technical Cooperation
ha	hectare
IBAMA	Brazil Environment and Natural Resources Institute
IDB	International Development Bank
IFAD	International Fund for Agricultural Development
IFF	Intergovernmental Forum on Forests
IMF	International Monetary Fund
INAB	National Forest Institute (Guatemala)

IPAM	Institute of Environmental Research for Amazônia
IPF	Intergovernmental Panel on Forests
IPRA	Indigenous Peoples Rights Act (the Philippines)
ITTO	International Tropical Timber Organization
IUCN	World Conservation Union
JFM	joint forest management
km	kilometre
LASA	Latin American Studies Association
m	metre
MARN	Ministry of the Environment and Natural Resources (Guatemala)
MDG	Millennium Development Goal
MIRNA	Integrated Management of Natural Resources in the Western Highlands project
MPR	People's Consultative Assembly (Indonesia)
NGO	non-governmental organization
NIPAS	National Integrated Protected Area Systems Act (Philippines)
OECD	Organisation for Economic Co-operation and Development
PARPA	Prioritization of Strategic Areas in the Western Highlands project
PROFOR	Forestry Programme of the World Bank
PVO	private voluntary organization
RSPB	Royal Society for the Protection of Birds
SAGEDEDOM	Service d'Appui à la Gestion et au Dévelopment des Communes (Burkino Faso)
SDC	Swiss Agency for Development and Cooperation
Sida	Swedish International Development Cooperation Agency
SRA	social responsibility agreement
TREASURE	Timber, Recreation, Environment, Aesthetics from a Sustained Usable Resource programme
UCA	Central American University
UK	United Kingdom
UN	United Nations
UNCD	United Nations Capital Development Fund
UNDCP	United Nations International Drug Control Program (Bolivia)
UNDP	United Nations Development Programme
UNFCCC	United Nations Framework Convention on Climate Change
UNFF	United Nations Forum on Forests
US	United States
USAID	US Agency for International Development
USDA	US Department of Agriculture
USSR	Union of Soviet Socialist Republics
WRI	World Resources Institute
WWF	World Wide Fund for Nature
Z$	Zimbabwean dollars

Introduction

Jürgen Blaser, Christian Küchli, Carol J. Pierce Colfer
and Doris Capistrano

Background to a global exchange

Decentralization processes are taking place in more than 60 countries worldwide. These processes vary by sector, by the discretionary powers transferred to lower levels of governance, by the design and implementation of fiscal and other financial aspects and by degree of social responsibility. These processes are of central importance in political and economic change in all sectors of the economy, including the forest sector.

A few years ago, forestry decentralization was a non-issue for many countries. In the proposals for action of the Intergovernmental Panel on Forests (IPF, 1995–1997) and the Intergovernmental Forum on Forests (IFF, 1997–2000), decentralization is not explicitly mentioned, and it is only indirectly present in the recommendations on participation. Decentralization has become a theme in forestry only since substantial political changes have taken place in many countries. As a matter of fact, governance – of which decentralization is one of the most visible elements today – is a crucial issue in sustainable forest management. It is the quality of governance that may ultimately determine the fate of forest resources in all their aspects – economic, social and ecological.

Important changes in approaches to forests and people have led to remarkable gains in the application of good governance principles. The development of an international forest regime through the United Nations Forum on Forests (UNFF) and the work of the members of the Collaborative Partnership on Forests (CPF), particularly the Food and Agriculture Organization (FAO), International Tropical Timber Organization (ITTO) and the Center for International Forestry Research (CIFOR), with the World Bank in the driver's seat, have undoubtedly contributed to these changes. This international forest regime has provided a much-needed impetus for a re-examination of concepts on forest and people interactions, and has facilitated policy change in many countries. In many cases, it has helped to create legitimate spaces and recognition for local initiatives and longstanding experimentation on the ground. The combination of locally driven processes in concert with this international forest regime has led to significant changes in forest governance worldwide. For example:

- National forest programmes have become the focal point of the UNFF, placing the discussion of better forest governance at the country level.
- Criteria and indicators for sustainable forest management have been developed and will help to improve the monitoring of forest management.
- The advantages of linking buyers and sellers through the promotion of specific market mechanisms have been recognized.
- The gap between environmental organizations and those focusing on poverty reduction is, in some cases, narrowing.
- Multi-stakeholder involvement, debate and consultation have become the norm and have helped to increase transparency and accountability.
- Forest law enforcement and governance initiatives, as promoted by the World Bank in Asia and Africa, have opened the debate on illegal logging and associated trade and corruption – themes that had been excluded from any substantial discussion of sustainable forest management.
- Numerous countries have attempted to reorient forest management by promoting greater decentralization and devolution to local people.

The decentralization processes occurring around the world have achieved momentum. Local and regional perspectives and agendas are increasingly informing and enriching forest-related discussions at the global scale. However, more is required to build local involvement in the global dialogue. Reaching global goals pertaining to both forest management and human well-being requires policies that are more relevant locally, as well as greater institutional capacity at both national and sub-national levels.

The Interlaken workshop Decentralization, Federal Systems of Forestry and National Forest Programmes, was therefore a very timely event given the sequence of actions undertaken over the past few years to secure conservation and sustainable management of forest resources. Decentralization is a cross-cutting issue, relevant to all the different aspects of sustainable forest management, which links sustainability objectives at the local level with broad global goals as defined in the UN Millennium Development Goals (MDGs).

We hope that the contributions made at the Interlaken workshop will stimulate further work at this cutting edge of policy, as well as greater connections between such efforts and broader development concerns. Intergovernmental global processes such as the UNFF can play a critical role, not only in shaping the global agenda, but also in facilitating and supporting the search for appropriate local solutions through local initiatives. Rio 1992 has taught us to think globally; Interlaken is a point of departure to learn how to act locally.

The Interlaken workshop and its participants

The objectives of the Interlaken workshop were as follows:

- to analyse the implications of decentralization of forest management for the development of national forest programmes and to identify strategies that would allow such programmes to effectively address this issue;
- to share the experience of countries that have decentralized their forestry systems with countries currently undergoing rapid processes of decentralization, including those in transitional phases;
- to derive the lessons learned from countries that have implemented decentralization for use, where suitable, in other countries;
- to prepare reflections and proposals for the consideration of the UN Forum on Forests related to decentralization, centralized systems of forestry and their implications for national forest programmes.

About 160 people from 51 countries participated in the workshop, representing 70 per cent of the global forest area. Approximately 75 per cent of the participants came from developing countries and countries in economic transition; 32 participants came from non-governmental organizations (NGOs) and 32 from private-sector organizations. The Interlaken workshop was an expert meeting, and participants expressed their views in their personal capacities, not as country representatives or representatives of specific institutions.

Because the intent of the workshop was to share ideas and experience and contribute to our global understanding of the processes related to decentralization, the workshop was divided into formal presentations, facilitated discussions on pre-selected topics, field trips (discussed in Chapter 8) and working groups. The formal presentations ranged from thematic discussions of decentralization, to surveys of regions or governance types, to country-specific analyses. The decentralization implications at various scales were also addressed, with presentations from participants representing international, national, sub-national and local levels.

The presentations revealed substantial variation across the globe in the history of governance approaches, in the extent and depth of people's participation, and in the balance of power among different governmental levels. Although some fascinating patterns emerged, a recurring theme was the uniqueness of each case and the importance of taking contextual factors into account when considering new governance modes. These patterns and variations will be explored in more detail in the following chapters, with particular attention, in the final chapter of this book, to the lessons we can learn.

The working group sessions were organized around six main themes, and their results were incorporated within the report submitted to UNFF:

- allocation of roles and responsibilities, and coordination at different levels and across sectors;

- maintaining ecosystem functions, sustaining forest productivity and appropriate application of knowledge and technology;
- policy, regulatory frameworks and equitable benefit-sharing;
- financial incentives, promoting investment and private-sector partnership;
- participation, conflict and multi-stakeholder processes; and
- capacity-building and technical and information support.

Interlaken workshop participants were highly constructive and cooperative, leading to an unusually candid sharing of experience and perspectives. The field day, in which people could experience decentralization on the ground, was instrumental. Overall, the workshop succeeded in defining issues and approaches towards decentralization and in giving a broad overview of existing and planned processes of decentralized forest management. This book pulls together the central descriptive and analytical conclusions from this fruitful sharing of global experience.

Organization of the book

The remainder of this book is organized into three main parts. The first, focusing on thematic issues, raises important cross-cutting questions. The second part offers case studies that convey some of the breadth of experience of individual countries. The third, based on a community panel, provides a 'bottom-up' perspective, demonstrating how decentralization policies have played out in rural communities in three countries – Guatemala, the Philippines and Zimbabwe.

Part I begins with an overview of forest governance in federal systems by Hans M. Gregersen, Arnoldo Contreras-Hermosilla, Andy White and Lauren Phillips. A longer version of Chapter 1, which included a great deal of case material, was published in draft form and used at the Interlaken workshop as a discussion document. This chapter, of necessity, captures only the highlights of the authors' study.

The authors of Chapters 2 and 4, who have, in fact, worked together in the past, focus on the important components of effective and benign democratic decentralization, though both argue that such a process has hardly been attempted in any real sense. In Chapter 2, Anne M. Larson surveys experiences in Africa, Asia and Latin America, focusing on lessons learned; and in Chapter 4, Jesse Ribot is more prescriptive, analysing the mechanisms and 'excuses' used by central governments to water down decentralization efforts.

Chapter 3, by Ian Ferguson and Cherukat Chandrasekharan, switches to a regional perspective, surveying the decentralization experience in Asia and the Pacific. Like Larson and Ribot, these authors find many problems with the implementation of decentralization; but they seem to favour a greater role for central government in the overall balance among the levels.

Chapter 5 details the experience of the FAO with decentralization in the forest sector. Merilio G. Morell outlines the various programmes that have supported decentralization and then provides two case studies (Burkina Faso and Mali) from which he draws a number of conclusions.

Chapter 6, written by Jeffery Sayer and colleagues Christopher Elliott, Edmond Barrow, Steve Gretzinger, Stewart Maginnis, Thomas McShane and Gill Shepherd, focuses on the implications of biodiversity conservation in decentralized forest resource management. Although supporting the reasoning behind decentralization, these authors warn of possible dangers to biodiversity and resource conservation unless some important functions remain in the hands of the state. They conclude by proposing some conditions under which decentralization can favour biodiversity conservation.

The shortcomings of decentralization policies, in their implementation, are clearly outlined in all of these contributions. Some authors argue for slowing the pace in order to give governments and citizens a chance to adapt to the new features of a decentralized approach; others suggest that local governments and citizens will become adept at dealing with their new powers only by using them. Although all see the potential value of decentralization, some favour a stronger central role and others a stronger local role, in the balance of power. Some show more faith in communities' management abilities, some have less.

Part II, the country cases, begins with Indonesia, the co-host of the workshop together with Switzerland. Chapter 7, written by members of Indonesia's Ministry of Forestry, Wandojo Siswanto and Wahjudi Wardojo, outlines the various laws and regulations that frame decentralization in Indonesia, and discusses frankly the principal problems that have plagued the process, as well as governmental efforts to solve them, in a country that began formal decentralization only very recently.

Chapter 8 presents Switzerland's decentralization experience through a presentation and discussion of the four field trips undertaken during the workshop. It is written by Christian Küchli and Jürgen Blaser, and presents a historical perspective on relations among the various levels of governance, as well as key factors instigating shifts in the balance of powers and responsibilities from more decentralized to more centralized forms and back again.

In Chapter 9, Pablo Pacheco describes the decentralization process in Bolivia, under way since the mid 1990s, which has focused on the devolution of significant powers to municipalities. Bolivia has empowered indigenous groups by returning their traditional territories to them, has empowered private landowners by allowing them to develop management plans and log their forests, and has empowered previously illegal loggers by legalizing small, community-based logging companies. Although significant strides have been made towards devolving powers both to communities and to lower levels of the bureaucracy, serious problems – outlined clearly in this chapter – remain.

Chapter 10, by Steve Amooti Nsita, describes the Ugandan situation. This country has been through several cycles of decentralization and recentralization, culminating most recently in another decentralization phase. The fact that many of the problems reported in Indonesia mirror those in Uganda does not augur well for a speedy resolution of their shared problems, particularly regarding the balance of power between levels.

Ghana, described in Chapter 11 by Oppon Sasu, is unique among developing countries in the longevity of its decentralized government, which was first formally acknowledged in 1878. Like Uganda, however, Ghana has gone through different phases. The current decentralization phase began in 1988, when local government was given additional powers, and was strengthened again in 2003 with a formal decentralization plan. A central problem in Ghana is the unwillingness of central government agencies to relinquish authority, as intended, to the district assemblies. This chapter includes serious attention to the lessons learned in Ghana's decentralization process.

Turning to Europe, in Chapter 12, Bill Ritchie and Mandy Haggith examine the decentralization process in Scotland, which involved oscillation between top-down and bottom-up pressures. The establishment of the Scottish Parliament in 1999 and the transfer of control of Scotland's forest estate to the Scottish Executive were two top-down elements leading to greater local control. For their part, local people's organized efforts to gain access to land and forests led to a land reform act that gives them the right to own land, including woodlands, and to the establishment of more than 100 community woodland organizations.

In Chapter 13, Natalia V. Malysheva outlines Russia's long history of centralized forest management and its painful efforts to decentralize during recent years. Malysheva, a member of the forest bureaucracy, looks at the historical evidence and comes down firmly in favour of a strong central state role in forest management, arguing for the importance of Russia's forests to the global community (over 25 per cent of the world's standing volume of timber is in Russia) and the threats posed by decentralization as implemented to date.

Chapter 14, by Gerald A. Rose with Douglas W. MacCleery, Ted L. Lorensen, Gary Lettman, David C. Zumeta, Mike Carroll, Timothy C. Boyce and Bruce Springer, describes forest management in a country with a longstanding form of federal government: the US. During the 20th century, concerns over environmental stewardship led the federal government to control many aspects of land management, including public forest managers' dealings with local communities. Some states fall back on those laws and regulations; others impose stricter standards. Colleagues Ted L. Lorensen, Gary Lettman, David C. Zumeta, Mike Carroll, Timothy C. Boyce and Bruce Springer describe the approaches taken in Oregon, Minnesota and Alabama.

The cases demonstrate serious problems with the implementation of decentralization, but also, with one exception (Russia), a commitment to continue trying to make it work. Switzerland, whose decentralization history started more than 150 years ago, experienced problems during the first decades very similar to the ones that newly decentralized countries are reporting. The long time to sort out conflicts and optimize the cooperation of all the governance levels might be one reason that this country case study, together with the Scottish experience and, perhaps, Bolivia, appears more optimistic in an otherwise rather dismal record. Recurrent problems include conflict over the division of authority and resources between the various levels of government, problems controlling forest crime, historical oscillation regarding preferred levels of decentralization, difficulties realizing the empowerment of communities as intended by decentralization advocates and unwillingness of central governments to relinquish control and resources to lower levels of government.

The final part of this book is devoted to community voices. Each of the first three chapters in Part III was written as a joint contribution by a community member and a partner who helped with writing, language and adjusting to the context of an international workshop. Our intent was to make the presentation of community views to an international body of policy-makers, scientists and bureaucrats as open and seamless as possible.

Steven Hlambela is a community leader in Zimbabwe's Chiredzi District. He was assisted by Witness Kozanayi, a junior researcher working for CIFOR, who has experience in that community as well as others. Chapter 15 outlines this community's experience in trying to implement a community-inspired resettlement vision. After a series of difficulties, including internal conflict, outsiders claiming resources, and disagreements and inaction by government officials, the authors conclude that communities cannot 'go it alone'. Both bottom-up and top-down involvement will be necessary to accomplish community goals.

Adolino L. Saway Alyas Datu Makapukaw is a tribal leader from the Talaandig tribe in Mindanao (the Philippines) and Felix S. Mirasol Jr works for the Philippine Department of Environment and Natural Resources in Mindanao. Chapter 16 tells the story of their efforts to manage Mount Kitanglad Natural Park cooperatively. Although there have been conflicts and problems, the authors consider the decentralization process to be proceeding well in the Philippines, and to have had a positive overall impact.

Silvel Elías is originally a community member from a Guatemalan village but is currently a doctoral student at the University of Toulouse, France; Hannah Wittman is a doctoral student from Cornell University. These authors find serious problems with the decentralization process in Guatemala, presented in Chapter 17. Conflicts abound between a government that has traditionally ignored and abused indigenous rights, and communities intent on defending their rights. In some cases, by shifting governmental regulation to a more local level, decentralization actually causes a loss of indigenous control over natural resources.

The community examples share the experience of conflict between governmental entities and members of local communities, and among other stakeholders as well. But the authors vary in the degree to which they consider decentralization helpful. The Philippine authors, although acknowledging some problems, are basically optimistic that this process is beneficial and that problems can be ironed out; the Zimbabwe authors reluctantly conclude that they need the help of the government; and the authors of the Guatemalan case present conflict-ridden scenarios with the potential for adverse effects on local communities. Clearly, decentralization is having different impacts in different places.

The final chapter in this book, the conclusion, pulls together the important threads that emerged during the workshop and highlights interesting differences.

This chapter concludes by summarizing important definitions pertaining to the issues addressed in and used throughout this book. The definitions are based on those developed by Hans M. Gregersen for the World Bank (see www.worldbank.org/publicsector/decentralization/admindecen.htm).

TYPES OF DECENTRALIZATION

Political decentralization. Groups at different levels of government – central, sub-national (meso) and local – are empowered to make decisions related to what affects them.

Administrative decentralization. Different levels of government administer resources and matters that have been delegated to them, generally through a constitution. In terms of decentralization as a process of change, and according to the level of transfer of responsibilities, it is useful to distinguish between the following forms:

- *Deconcentration* redistributes decision-making authority and financial and management responsibility within the central government; there is no real transfer of authority between levels of government. Deconcentration may involve only a shift of responsibilities from federal forest service officials of the capital city to those stationed in provinces or districts.
- *Delegation* transfers responsibilities and authority to semi-autonomous entities that respond to the central government but are not totally controlled by it. Public forestry corporations and, in some cases, implementation units of some forestry projects – often donor supported – are examples of this form of decentralization.
- *Devolution* transfers specific decision-making powers from one level of government to another (from a lower level to a higher level of government, in the special case of federations) or from government to entities of the civil society. Regional or provincial governments, for example, become semi-autonomous and administer forest resources according to their own priorities and within clearly defined geographic jurisdictions. Most political decentralization is associated with devolution.

Fiscal decentralization. Previously concentrated powers to tax and generate revenues are dispersed to other levels of government. For example, local governments are given the power to raise and retain financial resources to fulfill their responsibilities.

Market decentralization. Government privatizes or deregulates private functions, as has happened in the New Zealand forest sector.

Part I

Themes in Decentralization

Chapter 1

Forest Governance in Federal Systems: An Overview of Experiences and Implications for Decentralization

Hans M. Gregersen, Arnoldo Contreras-Hermosilla, Andy White and Lauren Phillips

INTRODUCTION

The role of government has been the focus of great debate in recent years. Much of this debate has focused on the reality of reduced government, increased reliance on markets and on private initiative, as well as on the important contributions of civil society and the private sector in providing public services. At the same time, there has been widespread and active debate on the optimal roles of different levels of government: how government authorities and responsibilities should be distributed among different levels of government. A World Bank study in 1999 found that more than 80 per cent of all developing countries and countries with economies in transition are currently experimenting with some form of decentralization (Manor, 1999).

The forest sector has not escaped these trends. Internationally recognized problems such as illegal logging and uncontrolled deforestation are increasingly attributed to weak governance structures. These problems, as well as the broader political trends, are driving many countries to reconsider the role of government in administering their forest resources and others to move away from centralized systems of decision-making and direct government implementation of forest programmes.

Unfortunately, the flurry of debate and political activity has often not benefited from the careful analysis of broader experience. Despite all the experience and innovation across the globe, there have been relatively few attempts to understand how different levels of government interact and balance authority and responsibilities in the forest sector, and how local governments, the private sector and civil society affect progress towards improved management of forest resources.

In this context the experience of federal systems of government in administering forest resources is particularly valuable. Federal systems of government are composed of member states or provinces and thus have, by definition, decentralized systems of governance. Some responsibilities and authorities are vested with the central federal government, and some with state or provincial levels. In federal systems, in contrast to centralized systems, meso- and local-level governments are often well established, with longstanding political constituencies and various accountability mechanisms that enhance their performance. Most importantly, the meso levels of government have not only responsibilities but also real authority and legal rights because they are part of a federal system defined by a constitution.

We review the experiences of selected major forest countries with federal systems of government and derive lessons for policy actors considering future decentralization initiatives, whether through a federal system or through some other system of government. The study focuses on the federal governments of Australia, Brazil, Canada, India, Malaysia, Nigeria, Russia and the US. Bolivia, Indonesia and Nepal have undertaken major decentralization programmes and are thus also included, even though they do not have federal systems of government. These 11 countries account for more than 60 per cent of the world's forests.

Each of these countries adopted decentralized forest governance systems at a different point in history. Their combined experience presents both common threads and dramatic differences. Those countries that adopted federal systems of government early on have largely adjusted to the administrative demands of harmonizing the operation of central and sub-national levels of government; others are still struggling with the complexities of decentralized management. Some have been more successful in securing the benefits of decentralized systems of governance while minimizing the associated dangers and costs.

DEFINITIONS: FEDERAL GOVERNMENTS AND DECENTRALIZATION

Countries with federal systems of government share responsibilities and authority, generally through the provisions of a constitution, between the national-level central government and meso (state, provincial or regional) and local levels of government. Powers between these levels are divided and coordinated in such a way that each level enjoys a substantial amount of independence from each other. This implies the existence of a constitution describing the division of powers and a means for resolving disputes. Most importantly, in contrast to simple devolution of specific powers and responsibilities from central to lower levels of government, federations use the principle of *constitutional non-centralization* rather than decentralization (Olowu, 2001).

In other words, when independent states decide to create a federation and a federal system of government, they confer, generally through a constitution, certain specific responsibilities and authorities to the federal government in the interest of all states. All other powers, responsibilities and rights remain with the states. In contrast, unitary governments may have subnational levels of governments; but these are not constitutionally empowered to make decisions on major government services and functions; rather, they are subordinate units. Indeed, for these reasons, use of the term *decentralized* is somewhat awkward in the case of federal governments. In the US, Canada, Malaysia and Australia, for example, authority for forest administration was never centralized at the federal level. Because of this confusion, we use *decentralized* to refer to the non-centralized distribution of authorities and responsibilities. Other federal governments, notably Russia and India, began as centralized governments, later adopted federal constitutions, and have been 'decentralizing' authority and responsibilities.

In a federal system, the central government usually has responsibilities for those resources, activities and events that affect more than one state and that are involved in the production of national (and sometimes international) public goods associated with the environmental services derived from forests. The member states generally have responsibility for and oversight of those resources, activities and events that affect mainly the state in question, the regulation of private forest practice and enterprises, and those functions that depend heavily upon local participation and involvement. Often, the federal government influences or controls state activity through federal laws, incentives and checks and balances related to the use of resources. Member states, in turn, generally regulate and guide the actions of lower levels of government (municipalities and districts), local community entities, private individual landowners and private companies operating within the states.

Variations in federal systems of government are considerable, however. There are differences in the relationship between responsibility and authority at different levels of government within federations; there are differences in the distribution of fiscal responsibilities; and there are many other differences that distinguish various federal forms of government. Federal systems can be simultaneously decentralized in some respects and centralized in others, and, indeed, there is constant tension between different levels of government.

Forest governance in federal systems

This section briefly describes the current structure of forest administration in eight major forested countries with federal systems of government, and identifies particular patterns in the distribution of government authority.

Most of the countries in our review are undergoing transitions in their forest administration, and the roles, functions and orientations of forest agencies and forest management are in substantial flux. We often found disagreement or general lack of knowledge about the actual distribution of authority and responsibilities in many countries, and a wide discrepancy between the official and actual distribution of power. Our findings represent our understanding at the moment; but the situation is very fluid in many of the countries studied.

The following nine points highlight patterns of forest administration in the eight federal countries:

1 Federal structures of forest governance tend to be complex and multifaceted, with strong cross-sectoral linkages to agriculture, water, transportation and other sectors (see Broadhead, 2003; Dubé and Schmithüsen, 2003). In all cases the federal forest agency is only one of several federal agencies administering public forestlands. Strong roles of other agencies and linkages to other sectors appear to help to create checks and balances for accountability and to ensure that the forest sector reflects the concerns of stakeholders, particularly beyond those directly involved in the forestry sector. In some countries, the other sectors involved can number into the hundreds. In the US, for example, some 31 federal entities interact directly with the Forest Service in planning and managing federal forestlands, and many others have a more indirect linkage (Ellefson and Moulton, 2000).
2 With the exception of the US, in all federal countries examined in this review, governments own a majority of all forestlands. Interestingly, of these seven countries where public forest predominates, majority ownership rests with the federal governments in Brazil, Russia and India. In contrast, in Malaysia, Nigeria, Canada and Australia, it is the state or provincial level that owns the majority of all forestlands. Federal ownership is substantial even in the US, where the federal government owns about 35 per cent of all forests, the states own about 5 per cent and the private sector owns the majority – about 60 per cent.
3 Policies and government structures to deal with the private sector and the civil society vary widely. In the US, the size of the private sector is considerable; accordingly, federal as well as state governments have established regulations and programmes to encourage and regulate private enterprises. India, in contrast, denies private corporation access to public forests and induces corporations to establish partnerships with small 'non-forest' owners. In some countries, the access of non-governmental institutions to the government decision-making process is encouraged; but in others such linkage is not promoted actively.
4 In many countries, federal and/or state governments do not officially recognize traditional land ownership rights. Thus, they deal in different ways with the interactions between local populations and local governments, with profound implications for the sector's governance.
5 The degree of responsibility and authority for the forest sector vested in the federal government and other tiers of government varies widely. In some the administration of the forest sector is relatively centralized, while in others main responsibilities and authority reside either in the second or even third tiers of government. In Brazil, for example, until recently, most key decisions and implementation of programmes were under the aegis of

the Federal Environment Institute; in Malaysia, states enjoy a high degree of autonomy to design and implement their own programmes.

6 In Canada, Malaysia, the US, Australia and India, comparatively strong meso-level government forestry agencies dominate the picture, to some extent because there is little federal forestland and the functions given to the federal agencies are fewer. At the same time, federal entities hold major responsibility for trade, research, international relations in forestry and the establishment of environmental standards. In the US, Brazil and several other cases, there is more federal forestland ownership; thus, more management responsibility for public land resides within the federal agencies.

7 Federal forest agencies tend to be responsible for managing federal forestlands and providing overall leadership on forestry matters, but often have limited jurisdiction over the regulation of forest practice on private lands – a responsibility held, in most cases, by member states or provinces.

8 In Russia and Nigeria, where a majority of the forestland is owned by the central or federal government and managed by the central forest agency, central agencies are weak and control of public forestlands is fragile. India and Canada, where a majority of forestlands are owned by state or provincial governments, have a better record of effectively controlling the public forest estate. Thus, decentralized ownership of public lands appears an effective strategy, at least in some cases.

9 In most cases, the power of the forest administration agencies, whether federal or state/provincial, vis-à-vis other agencies of government, is relatively minor. Public forest administrations are often subsidiaries of ministries of environment or agriculture. In some cases the jurisdiction of forest agencies is shared with other powerful agencies, as in the US and Brazil. Management of inter-sectoral and inter-agency linkages is difficult and is not often achieved satisfactorily in most federal countries. Australia is an exception: the government administers forests based on a broad process of consultation and decision-making, involving various agencies and actors of the private sector. In most cases, federal structures do not ensure horizontal coordination between agencies of government. In Nigeria and India, this hampers administration of forest ecosystems that span local administrative boundaries.

Decentralized forest governance: Findings

The 11 countries studied present a rich array of history and experience in forest governance, and offer some general findings and lessons for those considering the decentralization of their governance structure. Below we discuss findings on:

1 the implementation of decentralization;

2. the role of forestry within the broader political context and the importance of cross-sectoral linkages;
3. the importance of ensuring adequate capacity, incentives and accountability; and
4. the importance of ensuring adequate participation by civil society and the private sector.

Implementation of political decentralization

- In most federal countries, decentralization processes involved sovereign states' assigning authority and responsibilities to a central government formed through a constitutional process. Exceptions include the Russian Federation, Bolivia and India, where decentralization efforts involved devolution from central to meso- and local-level governments.
- Even in countries where the central government owns most forestland (Russia, Nigeria, Nepal and Indonesia), the relative power of the federal forest public administration is low and forest agencies were generally incapable of influencing the main course of events. The forest sector was therefore a follower, more than a leader, in the decentralization process.
- The process of debating and adjusting the distribution of authorities and responsibilities is open ended. The ongoing tensions between different levels of government and political forces have often contributed to a better definition of governance responsibilities and authority at different levels, consequently reinforcing administrative checks and balances. Thus, decentralization processes can be seen as evolutionary, the balance of powers undergoing constant pressure and revision. In some cases, it is more revolutionary (for example, the former Soviet Union, Bolivia and Indonesia).
- What now appear effective and efficient decentralized systems took many years to achieve, with many adjustments to unforeseen events along the way. The present is a period of transition in countries such as Russia, Bolivia, Indonesia and Nigeria.
- The evolution of the distribution of forest administration authorities and responsibilities between central, meso and local levels of government has been part of much broader national processes of balancing authorities and powers in response to shifting goals, needs, resources and political processes.
- Decentralization of responsibilities and authority to the third level is generally difficult. First, these levels of government have rarely been vested with adequate authority, revenues and accountability mechanisms – and thus lack the capacity and political constituencies necessary to handle new responsibilities. Second, decentralization initiatives frequently assign responsibilities without the complementary rights or resources to motivate adequate performance. And third, second-level governments are sometimes inadequately prepared or are involved in mediating between the central and local governments.

- Decentralization initiatives in federal countries appear easier to conduct, and are more effective in the short run, than in non-federal countries. Decentralization initiatives in non-federal countries are more challenging because they necessarily entail developing local government capacity and setting new precedents for managing revenues and enforcing accountability. This experience suggests that policy-makers need to be careful in drawing lessons from decentralized governance in federal governments for application to non-federal governments.
- The objectives of the decentralization process were apparent in most countries; but the operational mechanisms needed to ensure a smooth transition were less clear.
- When administrative and technical human resources were scarce, urban issues with greater political visibility, such as health, education and transportation infrastructure, and agricultural demands tended to receive greater attention than the management of forests.
- Decentralization processes were often paralleled by a deconcentration of forest-related functions at the federal level. For example, in the US, devolution of forest administration authorities to the newly incorporated western states (and to the private sector through land grants) was paralleled by deconcentration of some Forest Service functions and decision-making to regional offices.
- Despite the trend towards decentralization of forest governance, today we see clear arguments for central or even international mechanisms (such as global conventions) to address the production of national or global public goods associated with the environmental services derived from forests.

Cross-sectoral roles and linkages

- In most of the 11 countries studied, many government agencies, in addition to the forestry agencies, are involved in decisions about forest resources.
- Cross-sectoral linkages with judiciary, agriculture, energy, transportation and environment are important in shaping approaches to forest governance. The complexities mount with the different responsibilities of agencies at different levels of government.
- Effective decentralization in the forest sector can occur only when functions of government in other sectors, such as taxation policy, law enforcement and political participation, are also subject to decentralization.
- Simultaneous and balanced fiscal, administrative and political decentralization – involving not only forest administration but also related sectors – is extremely difficult to achieve. Problems arise if a balance is not achieved.
- The degree and extent of decentralization varied during different periods in given countries. As mentioned above, the process is dynamic and depends upon political philosophies and government-wide adjustments

to the public administration in general, not only in the forest sector and forest-related agencies.

- National forest congresses with broad political support can help to bring different stakeholders together to shape a national vision for forests and a strategy that can be adopted and adapted by the sub-national levels of government, civil society, the private sector and the international community.

Capacity, resources, incentives and accountability

- Effective forest governance appears to depend more upon the capabilities of the managing entities than on any particular form or degree of decentralization or centralization of management functions. More specifically, such capabilities are required at all levels of government.
- Administratively and technically weak local governments often hamper monitoring and control of activities in forest reserves – which opens opportunities for local elite or private-sector domination. Thus, legal reforms are not enough: institution-building is also needed.
- Decentralization of administrative responsibilities without commensurate financial resources creates incentives to manage resources unsustainably and generate revenue to finance local government operations. Even if the central government provides funding, local governments' incentives to deplete forest resources may be intense if there is uncertainty over forest control.
- In Brazil, decentralizing financial resources without creating mechanisms to ensure responsible fiscal management led to local government unaccountability, indebtedness, disinterest in administering forest resources and lack of funds to meet national forest management priorities. The void also facilitated local capture by powerful economic and political interests for rent-seeking purposes.
- In Nigeria, state governments have only limited authority; responsibility and power lie with local governments, despite their lack of capacity to manage forest resources. The result often is turmoil and lack of progress in forest management.
- Decentralization in the forest sector in India, Indonesia, Bolivia and Brazil was generally characterized by initial scarcity of managerial and technical resources at the state and local levels. Some sub-national governments either ignored forests or used them unsustainably, leading, in some cases, to increased federal regulation for environmental protection.

Participation of civil society and the private sector

- The private sector and civil society play a crucial role in the success of governance of forest resources, not only in administering forest resources, but also in ensuring government accountability to civil- and private-sector concerns.
- In Brazil and Malaysia, state-level public administrations have been strengthened; but little power has percolated to local governments and potential partners in the civil society and the private sector.
- In Brazil, Malaysia, Nepal, Bolivia, Indonesia and the US, national and international non-governmental organizations (NGOs) have partnered with local entities to improve local governance and the protection of forests. These partnerships appear to have worked better where mechanisms existed for communication not only between levels of government, but also between government and other stakeholders.
- In the US, where the private sector owns the majority of the forestland, state forest agencies are often responsible for regulating the private sector, with a main focus on environmental impacts and fair business practices of commercial forest activities; the federal government is involved with the private sector through incentive and fiscal programmes. In Canada, Indonesia, Australia, Nepal and Bolivia, involvement with the private sector occurs through the granting and regulation of long-term concessions and other contracts for private use and the management of public lands.
- In nearly all countries, the governments at all levels have considerable interaction – sometimes contentious – with indigenous peoples who have claims on land and forest resources. Bolivia and Nepal provide positive examples, where governments have recognized community forest rights, vesting these stakeholders with strong incentives to protect and improve their forest resources.
- Experiences in the US, Canada, Australia, Nepal, Bolivia and India suggest that local citizen group participation prevents governments from imposing measures that conflict with local conditions and traditions. This is particularly important in the non-federal countries. Participation in decision-making may increase if advocacy groups can organize disadvantaged groups, highlight the costs of maintaining the status quo, and provide technical services (such as monitoring and dissemination of information).
- Often the drive towards more decentralized forest governance has been followed by shifts in the public–private balance, both in terms of management and forest ownership.
- Many of our findings demonstrate that building a successful forest governance system requires the input and consideration of agencies, civil society groups and other entities far beyond the narrow confines of forest agencies.

Implications for forest governance

A framework for decentralization

We suggest that the basic framework for successful decentralized forest governance entails three elements:

1. *Sharing decision-making authority and responsibilities for forest management, including revenue-sharing, between different levels of government*:
    - Is there adequate technical and political capacity to govern and make public-interest decisions at each level?
    - Do existing laws harmonize national, regional and local objectives and functions?
    - Do communication and support functions link the different levels of government?
    - What responsibilities are best carried out at each level, and which are best given to or shared with the private sector and civil society groups?
    - Do fiscal responsibilities coincide with management responsibilities, such that responsibilities can be adequately discharged?
    - Can local levels levy taxes and fees without double taxation?
    - Are forest revenues linked to budgets and expenditures at the different levels of government?

2. *Enforcement of accountability at all levels of government*:
    - Is local political competition allowed and encouraged by higher levels?
    - Is there transparency in government decisions and actions at all levels?
    - Are the different levels of government accountable to each other, and does each support the accountability needs of other levels?
    - Are there effective public checks and balances on power use and misuse?
    - Is public information adequate to ensure transparency for all stakeholders?

3. *Linkages with other government sectors and agencies and with the private sector*:
    - Do sectors share power at all levels of government, and are conflict-resolution options effective?
    - Are there effective relationships with the private sector and regulation of private activity?
    - Are there mechanisms to govern and support intra-governmental linkages and authority-sharing?

The relationship created between authority, accountability mechanisms, responsibilities and revenue-sharing at different levels of government determines the success of decentralized systems of forest governance. Power without accountability can lead to arrogance, blatant favouritism and corruption. Decentralized power without accountability can lead to exclusion of certain stakeholders and to local corruption and waste. And responsibility without authority – or responsibility without resources – leads to ineffectiveness: the government entity that controls finances also, de facto, controls the activities that depend upon those resources for implementation. Therefore, while it is essential that local governments have decision-making authority, responsibilities and resources, accountability is crucial.

The third element in the list – cross-sectoral linkages – relates to the complexity of forest governance in democratic systems with many actors and stakeholders, each with a different perspective. Stakeholders must have a forum and mechanism for making their interests known. And it is important that there are clear 'rules of the game' (allocation of ownership, responsibilities, resources and authority) and mechanisms to avoid overlaps, conflicts and inaction.

Forest agencies in federal systems

In what follows, we describe in summary form the decentralized approaches of eight nations.

Australia

- *Ownership*: 25 per cent private, 75 per cent sub-national levels of government. State governments (but not territorial governments) legislate forest practices and grant licences for forest management on public land.
- *Federal agency*: Department of Agriculture, Fisheries and Forestry.
- *Federal agency responsibilities*: environmental protection; management of the federal forest estate; control of international trade in forest products; land use in the territories; general directives for forest resources management for all states (overall sectoral strategies).
- *State agencies*: forestry is a small component of larger integrated departments or multi-resource agencies. State agencies are responsible for land tenure; land use; public forest management; water supply; and regulation of private forestland practices. See 'Ownership' above.
- *Coordination between levels*: national forest policy statement and regional forest agreements. The Australian Forestry Council has developed a set of national principles to be applied in the management of native forest resources used for wood production on both public and private lands; Natural Heritage Trust.
- *Regulation of private forests*: state governments apply the same codes of practices as on public forestland.

Brazil

- *Ownership*: natural forests are owned by government. The Environment and Natural Resources Institute (IBAMA) has main responsibility for managing forest resources.
- *Federal agency*: IBAMA, linked to the Ministry of Environment; this agency has regional offices. It centralizes national environmental issues and administers forests country-wide.
- *State agencies*: IBAMA can transfer forest management responsibilities to the states; but this seldom has happened in any significant scale.
- *Coordination between levels*: the Programa Nacional de Florestas is composed of projects designed with the participation of all levels of government and the civil society, as well as NGOs.
- *Regulation of private forests:* IBAMA regulates relationships with the private sector; but states also have authority to do so.

Canada

- *Ownership*: 23 per cent federal, 71 per cent provincial, 6 per cent private. Provincial governments legislate forest practices on provincially owned land and grant licences for forest management. The federal and provincial governments manage and regulate federal and provincial parks, respectively.
- *Federal agency*: Natural Resources; Canadian Forest Service. The agency handles international trade and relations; management of federal land; national reporting; aboriginal affairs; and national consensus-building. Responsibilities for environmental regulation, as well as science and technology, are shared with provincial governments.
- *State agencies*: most provinces have a ministry of forests responsible for managing provincial land; allocation of timber licences; forest policy legislation; and data collection.
- *Coordination between levels*: the Canadian Council of Forest Ministers is focused on making more effective and efficient linkages between federal and sub-national entities; a national forest strategy exists.
- *Regulation of private forests*: provincial governments legislate forest practices on private forestland.

India

- *Ownership*: 90 per cent federal, 10 per cent communities and private. Forest administration is jointly managed by the federal and state governments; but the states have the primary responsibility for implementation. States are expected to decentralize to third-tier levels of government.
- *Federal agency*: the Central Ministry of Environment and Forests issues national policy directives and has responsibility for other functions that

are national in scope, such as forestry research and monitoring of policy implementation.

- *State agencies*: state forest departments have primary operational duties. States also adopt state policies that must be in agreement with policies issued by the central government.
- *Coordination between levels*: the third tier of government relationships is left mainly to states. Joint forest management has been employed in some areas, which involves partnerships between government and communities. In joint forest management agreements, sub-state governments are responsible for forest resources planning.
- *Regulation of private forests*: industry cannot lease access to forest resources and thus does not obtain wood supplies from government forests. National forest policy encourages partnerships with local communities for securing forest raw material supplies. Some sectors of the paper industry remain protected by import tariffs.

Malaysia

- *Ownership*: forest ownership rests with the states. In only a very few cases do forests belong to customary communities.
- *Federal agency*: Ministry of Primary Industries. The National Forest Council coordinates planning, management and development of forest resources. Responsibilities include provision of advice and technical assistance to the states, as well as training and research facilities; control of functions related to trade, industrial development and the environment; and national forest policy formulation.
- *State agencies*: state forestry departments are responsible for land tenure and ownership; forest policy at state level; and implementation and monitoring of forest policies.
- *Coordination between levels*: the National Forestry Council is a forum for federal and state governments; together, the central and state governments have prepared national-level criteria and indicators for sustainable forest management, developed a domestic forest management certification system, and established a relatively robust forest regulatory and monitoring system, including the national forest policy.
- *Regulation of private forests*: state governments legislate forest practices within the broader context of the national forest policy.

Nigeria

- *Ownership*: all forest reserves (less than 10 per cent of the country's territory) are under the co-management of states and local governments, who award timber concessions. Constitutionally, the states have the greatest control over forest resources; but, in reality, the federal government exerts more power as it controls the money.

Forested lands outside of forest reserves are not put under any form of systematic management.

- *Federal agency*: Federal Department of Forestry, whose responsibilities include the formulation of national forest policy; land-use planning, forestry development and environmental management; projects of national interest; research, training and education; trade and industrial development; institutional development.
- *State agencies*: most states have state forestry departments. According to the constitution, states can develop their own forest policies within the framework of the national forest policy; but, as yet, none have done so.
- *Coordination between levels*: the national forest policy addresses coordination. Revenue sharing systems are in place between state and local governments. The national forest and wildlife policy is being reviewed to make its implementation more participatory; a national wildlife forest law is also being developed with the involvement of all stakeholders.
- *Regulation of private forests*: state governments can develop forest policies within the context of the national forest policy.

Russia

- *Ownership*: 92 per cent of forests are federally owned. Other forests may remain under ownership of the subjects of the federation or other public authorities. The federal government can share ownership with the subjects.
- *Federal agency*: the Russian Federal Forest Service (changing to the Federal Forest Administration) is responsible for financing the protection, renewal and organization of forest use. As part of its privatization process, Russia is moving towards a system of user contracts on productive areas of the forest fund (forest estate); the unproductive areas will be directly managed by the Federal Forest Administration.
- *State agencies*: some authority has been delegated by the Federal Forest Administration through regional regulation-making bodies; but such regulations will be enforced by the Federal Forest Administration. *Leskhozes* are state enterprises (a type of regional forest service) that combine forest administration and forest management responsibilities. Subjects can make decisions regarding allocation of forest plots for leasing and free use.
- *Coordination between levels*: federal government has the primary responsibility for coordination; but institutional uncertainty prevents effective harmonization between levels of government.
- *Regulation of private forests*: four types of licences can be granted to the private sector so that forest operations and the implementation of management plans will be undertaken by forest users (licencees), while the central government or the subjects focus on strategic, regulatory and enforcement activities.

US

- *Ownership*: 35 per cent federal, 5 per cent states, 60 per cent private. The federal and state governments develop legislation for their lands, respectively.
- *Federal agency*: agencies of the US Department of Agriculture (Forest Service), the US Department of the Interior (Bureau of Land Management, National Park Service, Fish and Wildlife Service, and Bureau of Indian Affairs) and US Department of Defense have regulatory and management responsibilities for the land that they administer.
- *State agencies*: state forest departments regulate private industry and forest practice on private land, and enforce state laws. Policy, legislative and enforcement arrangements vary.
- *Coordination between levels*: there is no formal mechanism to coordinate forest policy across levels of government – causing some tension between federal and state levels. Some states see federal government influence as excessive. The management approaches of the several federal agencies that have jurisdiction over federal forestlands are often dissimilar.
- *Regulation of private forests*: state governments legislate forest policy for private forest management. Federal environmental legislation (Endangered Species Act, Clean Water Act) affects forest practice on private lands, with each state having the flexibility to design its regulatory framework to comply with national legislation.

Lessons from experience

Potential advantages

Decentralization reduces bureaucracy and reduces decision-making congestion at the centre. Decisions, particularly routine decisions, can be made more quickly. It also leads to institution-building and better understanding of conditions, needs and constraints at the local level. It facilitates information flows between local and central governments, and between government and the civil society and the private sector. It promotes partnerships with organizations outside the government. Local knowledge can be exploited more fully.

If decentralization leads to privatization, or to organizations that function as separate profit centres, decisions are more subject to the discipline of the market. At least some subsidized operations will be eliminated.

It is easier to involve local populations, particularly if requested actions are linked to benefit-sharing, and it gives them a greater sense of ownership. A larger share of benefits can remain in the localities and communities which generate them. There is potential for a greater coincidence between local traditions and informal rights, on one side, and formal norms imposed by the government, on the other.

Political meddling by central powers may be more difficult. Corruption may decrease if central government officials have less discretionary power and if, with greater levels of participation, decisions of local officials can be more closely scrutinized.

Potential dangers

Decentralization makes coordination, implementation and monitoring of national policies more difficult. Economies of scale in procurement and other activities may be lost. Big-picture national forest management may suffer if decisions conditioned by local objectives do not coincide with national objectives, and decentralized organizations may have limited technical and managerial knowledge. Decision-making may be less transparent and less responsive.

Decentralization can result in the allocation of central resources to regions, ethnic groups or political associates, thereby threatening national coherence. The central government may lose revenues and person power. Decentralization may increase arbitrariness and corruption.

Decisions may be conditioned by purely financial considerations that may not coincide with national or even local objectives. Decisions may be socially or environmentally undesirable or unsustainable. Non-commercial objectives of national policy may be lost.

Local elites may control decentralized institutions and use them for their own benefit. Central government political meddling may be simply replaced by local political interference.

If local governments do not produce a substantial economic surplus, net transfers from the central government may be lost. Local ownership may be lost if benefit-sharing becomes more inequitable.

There may be overwhelming pressure to 'mine' the forest for immediate local benefit. If formal norms were previously ineffective and de facto informal norms prevailed, decentralization may increase conflict between formal and informal norms.

Decentralization may be a vehicle for central political parties to penetrate the rural and forest economy. If monitoring and control from the centre are loosened, particularly if decentralization is not accompanied by participation, there may be more opportunities for corruption. Local elites and local government officials may conspire more effectively to carry out corrupt schemes.

Policy options

Identify which national policies should override the preferences of decentralized bodies and establish clear rules for their enforcement at national level, including penalties and rewards. Implement programmes to promote national policy objectives. Ensure broad-based public discussion of forest decentralization policy issues by disseminating information and establishing consultative mechanisms involving stakeholders.

Promote training to strengthen local technical and managerial capacity. Promote decentralization selectively – only to those local institutions that are able to deal with the managerial and technical demands of the process. Tap local knowledge.

Use subsidies and other incentives to induce decentralized and privatized institutions to conform to national priorities, including environmental quality and social improvement.

Establish clear rules of the game so that local space for decisions will be constrained within the boundaries of national priorities. Identify stakeholders and promote democratization of decision-making. Ensure equitable sharing of benefits and costs at the local level. Promote transparency in decision-making. Create channels for community participation. Promote open and public procurement schemes.

Net economic transfers from the central government may be desirable on social and environmental grounds. Establish clear policies for subsidies to local institutions.

Encourage participation in decision-making, a free flow of information and transparency. At the local level, focus on establishing mechanisms for greater transparency. The loss of central government resources is an unavoidable cost of decentralization; but policies to promote training and capacity-building programmes may alleviate the tensions.

Establish forest management minimum standards for decentralized institutions. Good policy-making depends upon a thorough understanding of local conditions. If local norms lead to sustainable livelihoods and sustainable forest management, complete devolution may be appropriate; if not, create incentives.

Design accountability rules, including transparency for decision-making processes and penalties for misuse of local resources for political gain. Establish an independent monitoring office to which 'whistle blowers' can give information. Encourage the media to report suspected malfeasance. Diversify service providers for increased choice. Increase technical assistance to foster public–private partnerships.

Conclusion

Decentralization offers great opportunities for improved forest management, but also great challenges. It is far from being a final solution to the ills of the forest sector because significant possible disadvantages and dangers threaten its potential benefits. Progress towards achieving the many potential benefits of decentralization faces daunting hazards.

We have attempted to determine what factors make decentralization work in the forest sector in certain nations with a federal system of government, and in countries that do not have federal systems of government, but are operating with decentralized forest governance systems. The findings offer models and lessons for countries that are pursuing decentralized forest governance.

In all cases studied, effective and efficient decentralization took many years to achieve, with adjustments to many unforeseen events. Russia and Nigeria are in transition, for example, and the future of forests in such countries depends heavily upon building effective forest governance capacity in both meso and lower levels of government. Politicians, high-level government officials and forest managers in the public and private sectors need the skill and willingness to avoid administrative chaos and financial crises, which can create fertile conditions for the capture of government for private ends.

Our review also provides evidence that transitions to more decentralized systems of forest governance are extremely complex, and we have barely touched on the intricacies. Although there is a vast and rich literature on decentralization and forest management and governance, much of it does not specifically address the incentives, the political economy and other factors that affect forest administration systems, or the capabilities and managerial and governance capacities of the entities within such systems. Although this book provides useful additions to our understanding, additional assessment is needed.

Note

This chapter is adapted from a much more detailed analysis and discussion. See Gregersen et al (2004).

The authors thank David Kaimowitz, Christian Küchli, Laurence von Schulthess and Adrian Sommer for their foresight in conceiving the workshop and for their invitation to prepare this analysis. As the project unfolded many individuals contributed ideas and materials, helping to ensure the accuracy of the country profiles. The authors would particularly like to recognize Kolade Adeyoju, Mike Apsey, Brian Belcher, Neil Byron, Kerstin Canby, Ben Cashore, Cherukat Chandrasekharan, Hosny El Lakany, Sebastiao Kengen, Jon Lindsay, Jag Maini, Ali Mekouar, Miguel Serediuk Milano, Salleh Mohd Nor, Augusta Molnar, Sten Nilsson, Labode Popoola, Jesse Ribot, Alaric Sample, Sara Scherr, Franz Schmithusen, Markku Simula and Jeffrey Vincent. The authors also thank three anonymous reviewers for their excellent and insightful comments.

References

Broadhead, J. (2003) *Cross-sectoral Policy Impacts in Forestry: Examples from Within and outside FAO*. Document CSPI/WP/02, FAO Forestry Department, FAO, Rome

Dubé, Y. and Schmithüsen, F. (eds) (2003) 'Cross-sectoral policy impacts between forestry and other sectors: 'Policy and legal, economic, environmental accounting and social aspects'. FAO Forestry Paper No 142, Rome, Italy

Ellefson, P. and Moulton, R. (2000) 'Fragmentation of forest resource agencies and programs: Challenges facing state and federal governments in the United States', in *IUFRO Sub-plenary Sessions*, vol 1, XXI IUFRO, World Congress, 7–12 August, Kuala Lumpur, Malaysia

Gregersen, H., Contreras-Hermosilla, A., White, A. and Phillips, L. (2004) *Forest*

Governance in Federal Systems: An Overview of Experiences and Implications for Decentralization. IFOR, Bogor, Indonesia

Manor, J. (1999) *The Political Economy of Democratic Decentralization*. World Bank, Washington, DC

Olowu, O. (2001) 'Decentralization policies and practices under structural adjustment and democratization in Africa', Democracy, Governance and Human Rights Programme, paper no 4, July, United Nations Research Institute for Social Development, Geneva

Chapter 2

Democratic Decentralization in the Forestry Sector: Lessons Learned from Africa, Asia and Latin America

Anne M. Larson

Introduction

Local governments are increasingly important actors in forest management. More than 60 countries say that they are decentralizing some aspect of natural resources management (Agrawal, 2001). And even where forest-sector decentralization is not part of national policy, local governments and local people often manage local forest resources, with or without formal mandates.

Looking at roughly 20 countries in Africa, Asia and Latin America, this chapter identifies common patterns and extracts lessons that will allow us to better understand decentralization successes and failures. We begin by briefly discussing conceptual issues that shape our understanding of decentralization in theory and practice. The next section presents the lessons learned regarding legal and institutional frameworks; financial management; the role of central governments; the role of third-party actors such as donors and non-governmental organizations (NGOs); local peoples; local governments; the timber industry; other factors affecting forestry decentralizations; and outcomes. The final section makes some suggestions for the future.

The issue of decentralization

Many of the studies discussed are based on the Center for International Forestry Research's (CIFOR's) study of the role of local governments in forestry management (Latin America and Indonesia), the role of local resource users (India) and the effects of national devolution – defined as the transfer of 'natural resource management to local individuals and institutions located within and outside of government' (Edmunds et al, 2003, p1) – on poor and marginalized groups, in particular (China and the Philippines). The primary resources for the Africa research were the World

Resources Institute (WRI) working papers on environmental governance. Another important source comprised the papers prepared for the 2002 Bellagio conference on Decentralization and Environment sponsored by WRI. Additional studies include other research from CIFOR and WRI, as well as from independent researchers. Taken together, the results of these studies paint a rich picture of the general and the specific, the common and the uncommon, for better understanding the dynamics of decentralization in the forestry sector.

Definitions of decentralization

Decentralization is usually defined as the transfer of powers from central government to lower levels in a political-administrative and territorial hierarchy (Crook and Manor, 1998; Agrawal and Ribot, 1999). This official power transfer can take two forms. Administrative decentralization, also known as deconcentration, refers to a transfer to lower-level central government authorities, or to other local authorities who are upwardly accountable to the central government (Ribot, 2002). In contrast, political or democratic decentralization refers to the transfer of authority to representative and downwardly accountable actors, such as elected local governments. To merit the term 'democratic decentralization', however, these representative and accountable local actors should have an autonomous, discretionary decision-making sphere with the power – and resources – to make decisions that are significant to the lives of local residents (Ribot, 2002). Democratic decentralization is often the yardstick against which power transfers are measured.

Decentralization is a tool for promoting development and is aimed at increasing efficiency, equity and democracy. Efficiency increases because greater local input should result in better-targeted policies and lower transaction costs (World Bank, 1997). As we will see, efficiency concerns are often the most important, in practice, to central governments. But the equity and democracy benefits (specifically, greater control over livelihoods and a greater share of other natural resource benefits; Edmunds et al, 2003) are likely more important to most local peoples. These are expected to come about by bringing government 'closer to the people' and increasing local participation, as well as government accountability (World Bank, 1988, 1997, 2000; Manor, 1999).

Most theorists now agree that local participation is essential for effective and sustainable natural resources management. Because of the failure to integrate local livelihood needs into outside interventions, for example, integrated rural development projects were often ineffective (Lutz and Caldecott, 1996), and many protected area projects actually increased biodiversity losses as well as social conflict (Enters and Anderson, 1999). National governments are often unable to control the sometimes vast forest areas under their legal authority (Carney and Farrington, 1998). And local people often ignore or filter rules imposed from outside; under the right circumstances, they are much more likely to respect rules that they had some role in creating (Gibson et al, 2000; Agrawal 2002). Hence, in theory, the institutional framework of democratic decentralization should also provide the conditions to enhance resource sustainability.

Why decentralize?

Decentralization reforms in the forestry sector are undertaken for various reasons. An important argument for decentralization relates to the historical exclusion of many local people from access to forest resources in order to promote commercial logging. As Edmunds et al (2003, p5) write: 'People living in forest areas ... have been expected to cope with sometimes drastic limitations on their choices and to yield rights of self-determination commonly enjoyed by others living outside of forests.' This applies to exclusion from protected areas, as well as from the economic benefits of commercial logging (among the countries studied, examples include, but are not limited to, Indonesia, Cameroon, Ghana, Mali and Bolivia).

In some cases decentralization policies have begun to right these wrongs, but only because national crises and vehement local demands forced the central governments' hand (deGrassi, 2003; Kassibo, 2003; Oyono, 2004; Resosudarmo, 2004). And in no case was righting these wrongs the primary reason that central governments adopted decentralization rhetoric or policy reforms.

The most common goal of decentralization in practice is to reduce costs (Colfer, 2005), often while increasing forest department revenues (Muhereza, 2003; Pacheco 2003), reaffirming private property rights (Beneria-Surkin, 2003; Pacheco, 2003) and/or addressing central government problems of legitimacy or economic and political crises (Bazaara, 2003; deGrassi, 2003; Kassibo, 2003; Oyono, 2004; Resosudarmo, 2004) at the same time. Democratization may be a stated goal; but in reality it is sometimes no more than official rhetoric. In fact, some studies found that decentralization policies actually served as a way of *increasing* state control over forest management (Wittman, 2002; Contreras, 2003; Sarin et al, 2003; see Chapter 17).

Donor pressure often plays an important role in initiating decentralization, though specific pressure in the forestry sector is less common than with regard to the service sector. Agrawal (2002) found that decentralization most often occurs when there is significant elite support within government, pressure from international donors (with financial incentives) and demands from local actors. Local actor demands may not be needed to initiate decentralization, he argues, but are necessary to actually bring about real political changes (see also Larson, 2004). Donor pressure, however, may result in 'imported' laws that are not appropriate for the national reality (Kassibo, 2003). And decentralization rhetoric may simply be employed principally to appease donors and to garner funds without actually implementing any meaningful changes (deGrassi, 2003).

As should be clear from the discussion so far, decentralization has different meanings for different people. These differences often go un-stated or even unidentified, and yet are essential for understanding the roots of problems and conflicts in practice, as well as for developing a basis for consensus. In particular, cost-cutting and improving conditions for private investment suggest a very different conceptual framework than securing local livelihoods and building a civic culture for democracy. While the two are not necessarily incompatible (with a clear understanding of the differences, and through extensive negotiation and compromise), in practice they suggest very different political, economic and social priorities.

The considerable obstacles to implementing meaningful decentralization in practice help to explain the many stops and starts of the process, its contradictions and highly varied outcomes.

LESSONS LEARNED

In this section we draw out the lessons learned from specific experiences without necessarily making conclusive judgements about the whole process at a national scale. The lessons tell us far more about problems with the *implementation* of democratic decentralization than about the effects or outcomes of democratic decentralization itself. This is because, in reality, what researchers have found is not democratic decentralization but partial, blocked and hybrid decentralization, and even centralization.

Legal and institutional frameworks

The transfer to local governments of significant authority regarding forest resources is rare

National governments often assert the importance of establishing autonomous local decision-making for the management of natural resources within each local jurisdiction. Yet, specific laws governing these resources, and forests in particular, limit this decision-making to playing, at best, a support role to the central government authority, or mandate coordination with the central government without establishing any specific mechanisms to do this. The result is a legal framework with contradictory laws that establish certain general authority locally, but then deny it when it comes to specifics (Larson, 2003d), laws on paper that are simply not implemented (Kassibo, 2003) or a discourse that is not implemented in practice (deGrassi, 2003). But without discretionary powers, local authorities have no legitimacy and are largely irrelevant (Kassibo, 2003).

In many cases the central government has outsourced costs while maintaining control (Edmunds et al, 2003). When some powers are, in fact, transferred to the local sphere, these usually involve responsibility without authority (Contreras, 2003; Ferroukhi and Echeverría, 2003; Kassibo, 2003;

Larson, 2003b; Pacheco, 2003), such as controlling crime and informal markets or carrying out tasks delegated by the central government. In Mali, the central government maintains control by delaying the implementation of key related activities, such as land delimitation, necessary prior to power transfer (Kassibo, 2002, 2003). Such delays have also been an important obstacle in Bolivia (Pacheco, 2004). Where some powers are transferred, the funds to carry out these powers are often not (Bazaara, 2003; Beneria-Surkin, 2003; Larson, 2003b). In Nicaragua as well as other countries, local authorities complain that they have been given the burdens, but not the benefits of natural resources management, either in terms of discretionary decision-making authority or income (Larson, 2003b).

Central governments also commonly maintain control over forest management through extensive bureaucratic procedures, such as forest management plans, price controls, marketing and permits for cutting, transport and processing (Colfer, 2005; Edmunds et al, 2003). In some cases this represents a loss of local decision-making authority, subjecting previously more autonomous organizations, such as India's *panchayats*, to central oversight and discipline (Edmunds et al, 2003).

Forest management plans are presumably an attempt to guarantee sustainable forest management; but there appears to be little monitoring of their implementation. Though technical arguments are given in defence of these plans, central authorities make not only technical but also *political* decisions, such as who has access to forest resources and how the income generated from them should be distributed (Bazaara, 2003). Some researchers have proposed alternatives such as setting minimum standards by region, within which local authorities have discretionary authority (Ribot, 2002). Elected forest councils in Kumaon, India, have historically operated under such an arrangement. They have an important autonomous decision-making space within rules and limits established by the central government, such as prohibitions on clear-felling, limits on the amount of fines and the need to get permission for the sale of certain products. The upward, as well as downward, accountability of the councils has been important to their success (Agrawal, 2002).

Only rarely are rights over decisions regarding *commercial* forest resources transferred to the local level, though some portion of the central government's income from taxes and royalties is now often returned to the local sphere (for example, in the Honduras, 1–10 per cent; Bolivia, 25 per cent; Nicaragua, 25 per cent; Guatemala, 50 per cent; Indonesia, 15–64 per cent). In eight case studies in four South-East Asian countries, central governments retained control over the most lucrative resources, leaving locals only subsistence uses, and thus precluding experiments in sustainable livelihood options. As the authors point out, however, to decrease poverty, people need resource access (Dupar and Badenoch, 2002). Nevertheless, forests are usually only devolved to local arenas after they are degraded (Colfer, 2005). And in at least one case, the state has re-appropriated forest resources after locals have invested in their protection and improved their degraded status, once they had increased in value (Sarin et al, 2003).

There are, however, exceptions. In the Indian state of Madhya Pradesh, well-stocked forests were also made available for joint forest management (JFM) (Sarin et al, 2003). In the state of Acre in Brazil, rubber-tappers obtained access to high-quality forests, though their distance from markets – making them less commercially attractive – combined with the tappers' organized pressure probably made this possible (Colfer, 2005). In Honduras, local governments can log their *ejidos*, which account for approximately 30 per cent of the country's forests, pending forestry institute approval of a management plan; nevertheless, there is a perception that *ejido* forests are more degraded than national forests (Kaimowitz, 1998).

Local governments in Indonesia would appear to have the most substantial control over commercial forests of all of the countries studied. In 1999, local governments were granted the right to authorize one-year small-scale logging permits to nationals for up to 100 hectare (ha) areas, though this right was then suspended (many local governments continued to grant permits anyway); a 2000 decree established the right of district leaders to grant permits up to 50,000ha (Resosudarmo, 2004). At the same time, the combination of laws and decrees is quite confusing and contradictory. Researchers argue that the forest department has sought to maintain control through various loopholes, such as excluding the substantial area already licensed, maintaining the right to control all pending applications and promoting a controversial plan to make large areas available to state-owned companies (Resosudarmo and Dermawan, 2002). At the same time, the state establishes all criteria and standards: classifying forests, and setting criteria and standards for all forest tariffs, forest products and forest areas (Resosudarmo, 2004; see also Chapter 7).

Whether local governments hold legal powers over local forest management or not, they often make decisions regarding forest resources. This may include imposing taxes or fees in violation of the law, authorizing or prohibiting logging, promoting reforestation or simply looking the other way with regard to violations. This does not mean that laws are not important. In fact, the failure to uphold the law may relate directly to the omission of local peoples from receiving benefits legally. Furthermore, legal authority to manage forests may generate its own dynamic. In several countries, such as Indonesia and Mali, local governments have taken initiatives based on their expectations of future legal rights (Kassibo, 2002; Resosudarmo, 2004). In addition, NGOs and donors may begin to direct investments towards building local capacity in accord with new legal competencies (Larson, 2003d).

Box 2.1 *Examples of decentralized forest management*

In Uganda, local governments may have significant powers but only over a very small area. Legal authority is complicated. Central governments control conservation areas and commercial logging in forest reserves of over 100ha and of trees on all private and public lands (Bazaara, 2003). Local governments are in charge of monitoring and vigilance; but there is little clarity regarding sanctions, arbitration or who should enforce what (Bazaara, 2003). In one district studied, of 18 forest reserves, 14 were privatized, 3 were recentralized and only 1 (49ha in extent) remained under local government control (Muhereza, 2003; see Chapter 10).

In Mali, as in Uganda, the Forest Service transfers many powers over some forests to local authorities. Nevertheless, these local domain forests have not yet been specified. The process, as outlined by proposed laws that have not been implemented, allows all higher levels of government to receive forest allocations first, leaving only those of little interest, and therefore of little value, for local governments. Hence, local governments may have extensive powers over few if any valuable resources (Agrawal and Ribot, 1999; Ribot, pers comm, 16 January 2004).

In Honduras, on paper local governments are given exclusive control over *ejidos* as their rightful owners (with logging subject to the national authority's approval of management plans); but they are only rarely able to exercise this right. Various additional rights are also granted by law; but in practice they have no authority whatsoever over the other forests in their jurisdiction (Ferroukhi and Echeverría, 2003; Larson, 2003d).

In Bolivia, local governments are given the right to 20 per cent of the national forest, in areas that are not currently claimed by anyone else or under concession. This area may then be assigned to local associations. Both the forest area and the associations must be approved by the central government, as well as by forest management plans. The process is long and bureaucratic (Pacheco 2003, 2004; see Chapter 9).

In Senegal, a new decentralized forestry law in 1998 gave local governments important rights to manage their forests (subject to approved management plans), to determine who could produce within their forests and to approve or veto proposed commercial production. The laws were to be fully implemented after a three-year transition period. Nevertheless, in 2003, the Forest Service was still making all major decisions, and rural councils apparently were not even consulted regarding the allocation of nationally approved logging permits. In addition, in Makacoulibantang, when the forestry officer requested the council president's approval for charcoal production, the president, as well as all the council presidents in the region, accepted in spite of widespread grassroots opposition. When pressed as to why, it became clear that they were 'beholden to the party authorities who select the candidates on the party slate and to the state with the power to allocate resources [that their] rural communit[ies are] in need of' (Ribot, pers comm, 16 January 2004).

Central governments can present compelling arguments for maintaining control

Three main arguments are heard repeatedly that defend retaining centralized control over forest resources: these involve the issue of scale and public goods; low local capacity; and the interference of politics. All are areas of concern; yet they are most often used to defend central interests and centralized policies, rather than to seek effective, negotiated decentralized alternatives.

deGrassi (2003) explores the way in which the first argument – forests as public goods – has been used to justify central control of forests over the past century. Central government, it is argued, is better placed to take into account scale effects, public service obligations and the protection of transboundary and transgenerational public goods, while districts will be tempted to realize the forests' cash value, and the short-term priorities of agriculturalists will prevail locally over long-term national interest. Nevertheless, proponents of this point of view fail to admit that central control has, in fact, been incapable of protecting forests (for more on central control see Chapter 11).

Local inadequacies are another common argument for maintaining central control over forest management. These concerns are often simply justifications for the status quo. As one author points out, the obvious response to low capacity is to *build* capacity, not to recentralize control (Bazaara, 2003); another approach is to retain certain technical decisions, but not *all* decisions. It is important to remember that most technical decisions also require certain social assumptions or judgements, and that technical experts have often failed to take into account what is sometimes more accurate local knowledge (see Klooster, 2002). Decisions that are 'technically correct' can also cause social hardships that may make them socially unacceptable.

In the third argument, forestry officials often refer to local governments as being 'too political' or to the need to 'keep politics out' of technical decisions (Pacheco, 2003). What they are referring to are problems such as patronage politics, political favouritism and the unwillingness to make hard decisions that may be technically necessary but socially undesirable. These are real problems; but they also plague central governments. In fact, local governments and other local actors are often powerless to act against irresponsible and destructive, but well-connected, logging companies with central government political connections.

Powers given by decree can be taken away

In Nicaragua a substantive plan for decentralization was reversed with a change in personnel at the forestry institute (Larson, 2003b). Ribot (2002) emphasizes the importance of making secure power transfers, arguing that people are unlikely to invest in rights or new institutions that they believe are insecure (Muhereza, 2003). Or they may urgently take advantage of their new powers, such as for logging, for fear that they may soon be withdrawn, resulting in over-cutting (Ribot, 2002). When powers are seen as privileges that can be withdrawn indiscriminately, the bargaining position of local people or governments who receive them is undermined (Muhereza, 2003); people fear taking the risk of making demands (Beneria-Surkin, 2003).

Authority is rarely transferred to representative, accountable local institutions

Decentralized powers are sometimes given to parallel committees set up by the central government for that purpose or to appointed personnel or traditional leaders, rather than to existing elected, representative bodies. Such policies may have the support of NGOs who distrust all representations of the state or their lack of capacity (Kassibo, 2003).

Sometimes called user groups or stakeholder committees, parallel committees are usually intended, at least by donors, to give local citizens greater direct influence over the development decisions that affect them (Manor, 2004). This does not appear to occur in practice, however. Many public officials use them to co-opt and control civil society; they are set up by and are accountable to central officials, and are given only limited powers (Sarin et al, 2003; Manor, 2004). Even when members are elected, government appointees may maintain control (Sarin et al, 2003; Oyono, 2003, 2004; Baviskar, 2004). Control may also be maintained through strict and opaque guidelines and bureaucratic procedures that only the government-appointed members can understand (Baviskar, 2004).

Community forestry committees in Cameroon were set up too fast and with inadequate guidelines and rules; they had no direction regarding what they must actually do or indicators of effective operation (Oyono, 2003, 2004). Rather than adapting local institutions that had much greater local legitimacy and authority, these committees were superimposed upon existing social systems.

The overall effect of such committees has been detrimental both to grassroots participation and to elected local governments (Sarin et al, 2003; Manor, 2004). Even when the parallel committees are elected, the proliferation of local-level committees can become overwhelming, and local people lose interest, particularly when committees have no real power and cannot respond to people's concerns (Mapedza and Mandondo, 2002). Manor (2004) argues that these committees should be placed under the control or strong influence of multipurpose, democratic institutions.

In several African countries, in particular, traditional leaders are sometimes selected to receive decentralized powers (Mapedza and Mandondo, 2002; Bazaara, 2003; Ntsebeza, 2004). Traditional leaders may be very important legitimate local institutions; but they also may be completely autocratic, unaccountable and undemocratic. They may even base their authority on fear and terror.

The legal framework for decentralized forestry is often confusing and contradictory

Laws may be confusing and contradictory, and it is often unclear which rules take precedence under what circumstances. Ambiguity leaves policies vulnerable to 'the politics of contradictory interpretation' (Sarin el al, 2003, p60). In Indonesia, local governments defend their rights based on one (decentralization) law, while the forestry department defends its actions based on another (forestry) law. Even a single law (decentralization) has serious contradictions within it regarding forestry authority; and the hierarchy of laws, regulations and decrees is unclear. These contradictions are due, in part, to insufficient preparation and planning, as well as apparently divergent interests within the central government (Resosudarmo and Dermawan, 2002; Resosudarmo, 2004; see Chapter 7). New laws intended to clarify contradictions sometimes just make things worse (Muhereza, 2003; Resosudarmo, 2004).

There is also often a general lack of knowledge about the law. Local governments may not know their rights and responsibilities, and citizens may be unaware of the responsibilities that their elected leaders have towards them. Legal ambiguities and contradictions make laws difficult to interpret even when the content is known and understood (Larson, 2003d). Kassibo (2003) argues that laws should be translated into simpler language, as well as into all national languages for broad distribution.

Effective institutions are needed for negotiating forest rights and interests

A few countries have established local environmental commissions (such as Nicaragua) or development committees (such as Guatemala) that can address forestry, as well as other local concerns. These multi-stakeholder committees usually involve actors from different levels of government and NGOs, as well as local citizens, operate under the authority of the local government and serve as a forum for negotiation that can then make recommendations to the appropriate authority for action. Institutions such as these, which are presumably more permanent, could help to increase continuity across changes of elected or appointed authorities (Larson, 2003d).

Financial management

Financial information is not always managed transparently

When forestry funds – royalties on logging and other resource exploitation contracts; park fees; reforestation fees; and various other specific taxes and levies – are managed by the central government for redistribution to the regions, local governments may not know what their share is supposed to be. Obtaining the full amount can mean a long battle (Larson 2003b), and they may not receive it in a timely fashion (deGrassi, 2003; Resosudarmo, 2004).

In Costa Rica, procedures for acquiring the local share of funds are complex, time consuming and bureaucratic (Ferroukhi and Aguilar, 2003). In Orissa state in India, the forest department simply failed to honour revenue-sharing agreements after the harvest; in Madhya Pradesh, locals were completely unaware of rules for budgetary allocations or the use of funds (Sarin et al, 2003). In Ghana, funds transfers are delayed, irregular, subject to central government approval and less than the amount needed and requested (deGrassi, 2003). In Indonesia, there are conflicts over the predetermined division of funds, ongoing calculations and transfer procedures, as well as substantial delays (Resosudarmo, 2004).

Those who manage funds should be accountable

The transparent management of funds, of course, would lead to greater accountability. Public events for the handover of forestry funds to community organizations had positive effects on downward accountability in Cameroon (Bigombe, 2003). Where there is some downward accountability, how funds are invested depends upon the social composition of the district or municipality and the relative influence of different actors (Pacheco, 2003). Upward accountability – not top-down management – can provide checks and balances on corruption, as well as undue influence of elite sectors.

There is sometimes little consensus on the appropriate distribution and use of funds

Rules regarding the distribution of forestry funds between the central government and the regional and local spheres can be set by central authorities or through negotiation among the different actors. Local actors often believe they deserve a larger share, especially if they struggled with forest protection for many years without government support (Sarin et al, 2003), if the local share is still extremely low (such as 1 per cent in Honduras, Vallejo, 2003) and if locals have historically seen central government or elites take everything (Bigombé, 2003; Resosudarmo, 2004). If the division of funds is not seen as fair, local actors have an incentive to try to redress this situation by gaining income from forest resources in other ways – by illegal logging, as in Cameroon and Ghana (Bigombé, 2003; Colfer, 2005); by granting more legal logging contracts, as in Cameroon and Indonesia (Bigombé, 2003; Resosudarmo, 2004); or by charging both legal and illegal fees, as in Nicaragua (Larson, 2003b, 2003c).

There is also disagreement regarding how such funds should be spent at the local level. The most urgent local priorities are usually not forest related; rather, the emphasis tends to be urban and on infrastructure (Pacheco, 2003).

Both the distribution of funds and their use should be subject to debate and negotiation through specific institutional mechanisms.

Distribution of funds to the local level may decrease some inequities but increase others

Regions where logging takes place stand to benefit from local control of forest income. At the same time, however, districts with fewer or lower-quality forests will probably now find themselves at a greater disadvantage. In Bolivia, for example, 80 per cent of forest income goes to only 30 out of 109 forested municipalities (Pacheco, 2003).

Reforestation funds could be used to redress such inequalities if they are made available on a national scale. Similarly, environmental services payments could redress inequities in districts where logging is prohibited in protected areas.

Central governments

Central governments often block or manipulate decentralization

Central governments may seek to maintain control in decentralization processes by denying the legal transfer of power to local governments. Or one office may promote decentralization while another obstructs it. Top-level forestry officials may also promote it, while lower-level officials try to block it. Forestry institutes sometimes promote centralization because civil servants want to maintain their power and income base (Bazaara, 2003; Larson, 2003a).

As Muhereza (2003) writes, understanding decentralization necessitates understanding the *interests of the state*. The distribution of powers over natural resources is more conflictive than the sectors that are most commonly targeted for decentralization – services and infrastructure – because forests are sources both of livelihoods and of wealth (Kaimowitz and Ribot, 2002; Larson, 2003a). Forest resources are a source of tributes of all kinds (Kassibo, 2003), as well as national revenue (Bazaara, 2003). It is not surprising that central governments resist giving up powers that they believe go against their own political as well as personal interests (Muhereza, 2003; Bigombé, 2003). Nevertheless, the 'public goods' interest of the state and the legitimacy of state-sponsored devolution arrangements are often taken for granted (Edmunds et al, 2003).

Central government oversight is important and necessary

Although too much oversight of local governments can be detrimental, checks and balances on local authority over forests are essential for good governance and to protect resources. Indonesia demonstrates the dangers that can arise from an abrupt loss of central authority. Local governments have been given important authority over forests after 30 years of marginalization, during which time the central government 'stole' local resources (in the eyes of the Outer Islands) and brought about a severe economic, political and social crisis. Local governments, which are not particularly accountable either to provincial or central government or to their citizens, seem determined to get from forests what they can while asserting their autonomy and economic independence from central government (Resosudarmo, 2003).

Central government authority, then, is necessary to address problems such as corruption, forest clearing and the externalities of local practices. Mandondo and Mapedza (2003) also argue that central governments can play a crucial supportive role and can help to overcome mid-level resistance to decentralization.

Forest department support can help make decentralization work

Open communication and exchange of information between the levels of government can help to reduce conflicts and build mutual respect and local capacity. Forestry offices in Ghana, Nepal and the Philippines provide important information to communities (Colfer, 2005). The 'Community Forestry Project', BOSCOM, in Guatemala has trained officials of local government forestry offices and helped to create a dialogue between the national and local level (Larson, 2003c). The lack of dialogue promotes the dispersion of efforts (Pacheco, 2003).

Sarin et al (2003) point out that forest departments could provide technical guidance, handle offenders and facilitate the resolution of boundary issues. In India, however, they more often disrupt effective arrangements (Sarin et al, 2003). A forestry department's failure to work with local governments may also arise from inadequate capacity and from corruption (Larson, 2003c; Oyono, 2003). Forest departments are also often criticized for dealing far too leniently with people accused of forest crimes that have been identified by locals (Sarin et al, 2003), hence undermining local initiatives to protect forests.

Forest departments generally have little interest in or understanding of 'the local'

Forest departments are often primarily interested in forest revenues and timber, whereas forest-dependent communities may have totally different priorities (Sarin et al, 2003). In India, for example, Sarin et al (2003) found that forestry officials were primarily interested in promoting projects that allowed them to control larger budgets, rather than cost-effective, community-based methods for improving forest quality. In fact, one method for increasing forest cover is still to keep forest-dependent locals out (Sarin et al, 2003). Forestry institutes pay little attention to the local forest management institutions that existed prior to decentralization arrangements – and even actively work to undermine them.

Third-party actors

Donor assistance can be crucial to making decentralization work

Donor pressure often appears to be an important impetus for getting central governments to implement some kind of decentralization (though this may only be superficial; de Grassi, 2003), and donor assistance can help to make it work. In Bolivia, the Honduras and Uganda, for example, donor assistance has enabled some local governments to exercise their powers effectively (Bazaara, 2003; Pacheco, 2003; Vallejo, 2003). In Bolivia, the local associations that have obtained approval of their forest management plans had help from the Bolivian Forestry project, BOLFOR. Though there are many problems with the NGO project SOS Sahel in Mali, locals are now well trained in forest management activities (Kassibo, 2003). In Cameroon, Oyono (2003) found that NGOs can be the key to negotiation and reconciliation.

Shackleton et al (2002) conclude that external actors can move policy and practice towards local interests. Similarly, in Brazil, Toni (2003) found that creative initiatives are more likely in Amazonian agricultural frontier areas where there are outside players such as NGOs and donor agencies. Where local people do not have the social capital or organizational capacity to make their needs heard, third-party actors can be important – for example, in defending the interests of marginalized groups.

If managed inappropriately, donor assistance can be detrimental

Donors sometimes have different agendas than the people with whom they are working, and thus undermine local initiative and empowerment; ignore or fail to recognize the divisive effects of their practices or their funding; and create dependence on outside financial support. Just as governments – central, state or local – cannot be expected to operate only in the public interest, NGOs and donors must also be seen as having their own organizational, as well as personal, interests.

Contreras (2003) found that outside assistance generated conflicts in the Philippines when the community vision of forest management differed from that of third parties (see also Chapter 16). In Mali, Kassibo (2003) found that third parties often promote one-size-fits-all development programmes that fail to take local realities into account.

In India, Sarin et al (2003) found that outside funding sometimes created jealousies, divisions and inequities among villages. It also generated conflicts within communities over the control of funds. Where such programmes required voluntary local contributions, the costs were disproportionately borne by poor women, while the funds were controlled by the male elite.

Donors can undermine sustainability by creating financial dependence. It is often easier for local groups to get NGO funds than depend upon their own resources, though the latter is more sustainable. Sarin et al (2003) argue that NGOs often encourage dependence to maintain control. In Bolivia,

Beneria-Surkin (2003) found that technical advisers sometimes failed to train locals, thereby prolonging their own jobs. In the Philippines, Contreras (2003, p130) warns of the formation of 'a whole industry of rent-seeking NGOs'.

Donors with an environmental bent sometimes have a strong conservationist bias

NGOs often fail to recognize local livelihood needs and support repressive environmental protection strategies (Bazaara, 2003; Kassibo, 2003; Sarin et al, 2003). The voice for conservation may be very strong precisely in places where marginal groups dominate, such as tropical forest regions, and the agendas for conservation, livelihoods and development do not always coincide (see Larson, 2001a). The history of the Chipko movement in India provides a perfect example of a movement for local control that outside activists mistakenly thought concurred with their own conservation interests (see Sarin et al, 2003). Nevertheless, the perspective of conservation NGOs may provide an important balance where deforestation is rampant.

Local peoples

Local forest management is often undermined through decentralization.

Before analysing new decentralization arrangements, it is important to understand what existed previously. Contributors to Edmunds and Wollenberg's (2003) collection found that decentralization strategies often impose new rules and authority over forest resources, either ignoring existing institutions or specifically attempting to undermine them (Contreras, 2003; Sarin et al, 2003).

In some parts of India, joint forest management brought areas previously managed autonomously under state control. New authorities were created, and unaccountable forestry department personnel were placed in charge of funds and all technical forestry decisions, based on the 'forestry department's vision of a good forest' (Sarin et al, 2003, p64). Existing local structures had been effective in terms of forest regeneration; but the new state-sponsored local institutions were sometimes less effective and eliminated incentives for forest protection.

In the Philippines, Contreras (2003, p130) found that central government policies were an attempt 'to colonize the spaces' that had already been opened by local initiative and take advantage of cheap, voluntary local labour. The new bureaucracy threatens organic structures that already exist.

In Bolivia and Nicaragua, some decentralization to local governments has occurred in tandem with giving indigenous groups greater control over their territories and resources. Where these groups hold important positions in local government, there may be fewer contradictions (Larson, 2001b; Pacheco, 2004); but further study is needed (see Larson, 2005). In

Guatemala, some fear that government attempts to control forest management on *ejidos* is aimed at controlling indigenous peoples (Wittman, 2002; see also Chapter 17).

Decentralization does not always make matters worse for locals, however. In Bolivia, local people previously had no legal right to forests and encroached on concession areas illegally; new provisions for legal access to forests appease both locals and concessionaires (Pacheco, 2004). And when local people do not accept decentralization policies, they may get officials to 'turn a blind eye' to local encroachment to maintain control (Sarin et al, 2003).

Decentralization rarely includes effective participation and accountability mechanisms

Mechanisms to promote grassroots participation and downward accountability are essential to democratic decentralization. For the local level, many countries mandate popular elections, participatory budget planning, fora for discussion and negotiation, periodic meetings between local authorities and local people, mandated financial reports and social audits or vigilance committees. Other accountability mechanisms include social pressure, the media, recall elections, central government oversight and popular protest.

If participation cannot lead to changes in policy or practice, however, it becomes meaningless. Social auditors in Cameroon have no recourse if access to information is blocked (Bigombé, 2003). Similarly, participation 'in projects, policies and laws that are predetermined by unrepresentative and unaccountable central powers' is not meaningful participation (deGrassi, 2003, p22). Conservation co-management projects cannot empower locals when partnerships are unequal and the project is top down – conceived of as 'participation for conservation' (Mandondo, 2000; Mapedza and Mandondo, 2002).

Bolivia has municipal participatory planning processes as well as community-based vigilance committees to oversee the local government's financial investments. These committees can request congress to freeze the local budget. In reality, though, planning is not particularly participatory, and plans are not usually implemented; vigilance committees do not often work and may be co-opted by political parties (Pacheco, 2004).

One important arena for participation is local political office. Some indigenous groups shy away from political participation, however, in part because running for office requires political party affiliation. To some groups, such as the Guarani in Bolivia, party politics is seen as meaning political favouritism and providing jobs to party members, and there is little cultural fit with the Guarani's decision-making traditions (Beneria-Surkin, 2003). Political positions are also frequently inaccessible to women.

Without representation, accountability or participation, local people find ways of asserting their claims through the failure to carry out instructions, manipulating rules, illegal logging or even burning down the forest. Mandondo and Mapedza (2003) argue that democratic decentralization should entail 'the forging of strategic compromises that firmly secure the interests of the most disadvantaged groups with outcomes regularly reviewed'. These reviews must include the power to reverse or realign governance arrangements to better meet intended goals. Who should do the reviewing and make decisions about these arrangements depends upon the specific circumstances. Accountability mechanisms should include legal recourse through an effective, unbiased and accessible court system that people understand (Dupar and Badenoch, 2002).

Local communities are differentiated

Central and local governments, NGOs and donors may fail to recognize local differentiation within communities. Marginalized groups may include women, youth, lower-caste groups, indigenous peoples and immigrants (Kassibo, 2003). Even when decentralization includes real local participation in decision-making, some groups may be left out. In Indonesia, for example, there are rising ethnic tensions over who represents 'the local', particularly older residents versus newer immigrants (Peluso, 2002; Resosudarmo and Dermawan, 2002). Many researchers argue that disadvantaged groups need direct involvement and empowerment, rather than indirect representation, unless specific mechanisms ensure the accountability of representatives (Edmunds et al, 2003). Promoting women's participation through activities to raise self-confidence and providing special channels for articulating concerns has had some success in north-eastern Cambodia (Dupar and Badenoch, 2002).

Local social capital and networks with outsiders are essential

Organized, effective local demand is essential to obtaining real power transfers (Agrawal, 2002; Muhereza, 2003; Toni, 2003; Baviskar, 2004; Larson, 2004; Pacheco, 2004) and to taking advantage of new opportunities that decentralization offers. In the Philippines, strong social capital, together with support from local governments and civil society organizations, led to the most effective implementation of forest devolution (Magno, 2001).

Shackleton et al (2002) found that organizational capacity and political capital led to better outcomes through resource mobilization and negotiation. Analysing eight case studies from South-East Asia, Dupar and Badenoch (2002, p54) found that 'clear demands for representation and good performance in government ... best exploit the potential of decentralization processes'. Social capital helped a community to defend its own interests when faced with state efforts to build a version of community forestry in the Philippines (Contreras, 2003).

Social capital can favour some local actors over others, however. In China, connections to government through family networks increase a household's ability to make demands; but families who do not have these ties are at a disadvantage (Dachang and Edmunds, 2003). In societies that are fragmented, decentralization programmes, such as community forestry in Cameroon, will suffer unless the problems of social capital are addressed (Oyono, 2003).

Local people and their leaders may accept policies detrimental to them and their resources if they lack information or are deliberately deceived (Sarin et al, 2003). Logging companies, in particular, are often able to manipulate local peoples (Resosudarmo and Dermawan, 2002; Resosudarmo, 2004). Communities who lack information and outside support in negotiating contracts with logging companies and timber brokers tend to receive a small share of benefits and are unable to enforce the contracts (Barr et al, 2001).

Effective networks with supportive outside actors could help local people to prevent such abuses. Federations of local organizations would also create a better support system at the village level, as well as a collective voice for negotiation (Sarin et al, 2003), although this scaling-up of local organizations can exacerbate the problems of differentiation and compromise downward accountability (Sarin et al, 2003).

Local people are more likely to follow their own rules

In Kumaon, India, Agrawal (2002) found that local people are self-regulatory, accepting the need for rules and their enforcement. Their acceptance is tied to their participation both in the rule-making process and in enforcement strategies. Colfer (2005) finds that where devolution is greater, local people tend to understand their formal rights and responsibilities more clearly.

Local governments

Elections may not ensure accountability at the local level

Throughout the case studies, there are complaints about electoral accountability. Most often, it appears that candidates are chosen by national political parties, and local candidates who are not party members are often ineligible to run. In order to improve representation and accountability, elections must be transparent, with third-party monitoring, credible oppositions, vibrant local media, active civil society and vigilant local citizens (Mandondo and Mapedza, 2003). In Kumaon, India, candidates are selected among residents and can be removed if people are unhappy with the way in which they are making decisions (Agrawal and Ribot, 1999). But in Bolivia, where mayors can be removed annually by the council for poor performance, their removal is often political and not based on performance at all (Pacheco, 2004). In one study in Indonesia, researchers found that 'almost no district government officials are accountable to district residents through direct elections, open meetings, public records, an independent press and the like' (Barr et al, 2001, p53).

Local governments may have little motivation to take forest-related initiatives

Most local governments' motivation to invest in forests tends to be low because they receive few benefits from forest resources; when they do, these initiatives tend to emphasize obtaining economic benefits.

In South-East Asia, Dupar and Badenoch (2002) found that citizens are more likely to choose local candidates who appear most likely to improve their material well-being. In many cases, 'development' has priority over 'environment', including forest management (Bazaara, 2003).

Local governments are often expected to incur costs for forest protection or monitoring from which they receive no compensation. They are unwilling to dedicate scarce financial resources to forestry when the majority of income from forests goes to the central government. In these cases they invest only funds from donors or special central government programmes (Bazaara, 2003). In Bolivia, local governments are more likely to control forest clearing than illegal logging, probably because they receive greater income from permits for the former (Pacheco, 2004).

Some local governments, however, have formed forestry or environmental offices despite a lack of mandates or funding, as in Bolivia (Pacheco, 2003). A few Costa Rican municipalities have also begun or maintained environment offices on their own initiative (Ferrouhki and Aguilar, 2003). In several Guatemalan municipalities, local people have demanded forest protection, and in one case they forced the local government to reopen its local forestry office (Alvarado, 29 July, 2003, pers comm).

Many local governments are primarily interested in the financial benefits of forest management. In the Philippines, for example, local governments are interested in political or fiscal benefits as a conduit for patronage relations (Contreras, 2003). Cash-strapped local governments in China, under new laws requiring them to raise their own revenue, encouraged some liquidation of forests for cash (Dupar and Badenoch, 2002).

Conservation areas that take lucrative forests out of commercial use may be opposed locally. Unless there are, at least, opportunities for tourist development, local governments may demand compensation (Resosudarmo, 2004). On the other hand, local governments may defend protected areas against outside interests in favour of locals, though they can also support outside interests if they think this will bring economic benefits (Larson, 2001b; Pacheco, 2003).

Capacity is commonly low among local governments but can be improved

'Capacity' – the ability of local governments to carry out their mandates – can include financial and administrative management, problem-solving, technical skills and the capacity to be democratic leaders. In forestry, 'capacity' often refers to the science of forest management, and sometimes financial management. Although forestry decisions require certain technical expertise, many forestry institutes reserve for themselves far more than technical decisions. Moreover, local knowledge may be superior to the technical expertise of forest professionals unfamiliar with local conditions.

In some cases, local governments have been effectively trained, or receive sufficient funding to hire trained personnel, to undertake forest management responsibilities (Ferroukhi and Echeverría, 2003; Pacheco, 2003). Funding can enhance capacity by facilitating the hiring of better-trained personnel and establishing forestry or environment offices with the personnel and resources to investigate forestry problems or conflicts, promote local projects, and so on. Urban municipalities with a larger tax base show clear advantages in this respect, as well as a few rural municipalities with significant NGO or donor funding (Larson, 2001b).

Local government forestry offices have often managed to improve coordination with central agencies and with local actors, provide technical assistance and become self-sustaining, using the income from forestry (Larson, 2003d; see Ferroukhi, 2003). But they may also be too dependent on foreign projects, under-funded and vulnerable to political changes.

Downward accountability is not sufficient to guarantee sustainable forest management

Local residents may not always want to conserve their forests. As a result, the downward accountability of local governments may lead to deforestation. In Uganda, candidates for local government have promised to open up parkland if they are elected (Bazaara, 2003). In another case, officials failed to enforce a ban on mahogany logging because local people wanted the wood to make school desks (Bazaara, 2003). Similarly, in some agricultural frontier areas that are dominated by new colonists, the majority of immigrants prefer clearing the forest for agriculture (Larson, 2001a; Toni, 2003). These realities suggest opportunities, under democratic decentralization, for seeking negotiated solutions that could work both for local people and for the forest.

Local government associations can improve the effectiveness of local government

The Nicaraguan Association of Municipal Governments (AMUNIC) has promoted legislation in defence of local government interests and sought responsible behaviour from local leaders. In Indonesia, local government associations have improved communication and strengthened bargaining power with other levels of government (Resosudarmo and Dermawan, 2002).

Regional associations allow local governments to address the management of environmental externalities and forests and water sources that span their borders. In the Honduras, there are almost 100 such associations, called *mancomunidades*, which confront common problems and improve their negotiating power. They invest jointly in projects such as local transformation of timber and seek better market opportunities (Larson, 2003d; Vallejo and Ferroukhi, 2005).

Implications for the timber industry

Decentralization may increase the timber industry's costs; but they may also find it easier to work with local officials

In Indonesia, industry representatives were anxious that central government permits would no longer be recognized locally and that they would lose their timber concession areas. In response, they invested in cultivating relations with the local elite (Resosudarmo and Dermawan, 2002).

In Nicaragua, logging companies were frustrated because the process was still controlled by the central government; but local officials now also demanded fees. The result was a new forestry law that outlaws these local charges. In Cameroon, timber companies admitted that they prefer working directly with decentralized structures than with administrative authorities and civil servants from the forestry ministry, 'whom they find hard to satisfy and always eager for more money' (Oyono, 2003).

Other factors affecting forestry decentralization

Macro-economic development goals may interfere with decentralized forest management

In Bolivia, the broader development pattern is privatization and resource exploitation, and local actors are unable to influence this (Beneria-Surkin, 2003). Similarly, in India, development is shaped by the government's desire to open up tribal areas to national and multinational investment (Sarin et al, 2003).

In Vietnam and Laos, production quotas and export targets of upwardly accountable line agencies do not necessarily match local priorities for subsistence (Dupar and Badenoch, 2002). In China, powerful central policy-makers undermine decentralized powers with decisions that affect people's livelihoods, such as a ban on logging in natural forests (Dachang and Edmunds, 2003; Xu and Ribot 2004).

Corruption, violence and power are detrimental to democratic decentralization

When there is corruption at the national level, decentralization may simply allow more local elites to participate in corrupt activities. Similarly, violence can interfere with the rule of law, as in Indonesia (Peluso, 2002). Actors who operate outside the law include traditional leaders in Uganda (Bazaara, 2003), the 'timber mafia' in Nicaragua (Larson, 2003b) and drug traffickers, who are suspected of occupying forested areas in Nicaragua (Pommier, undated) and Honduras (Vallejo, 2003).

Land tenure conflicts need to be resolved so that local forest management can be effective

In India, some legal ambiguities in decentralization policies are related to tenure ambiguity (Sarin et al, 2003, p60). In South-East Asia, Dupar and Badenoch (2002) concluded that the resolution of land tenure problems was essential for securing livelihoods, prior to effective decentralized development planning. Municipal reserve creation in Bolivia is limited by tenure conflicts and lack of information on the exact location of public forests (Pacheco, 2003). Resolving tenure issues is essential for improving the management of *ejidos* in Guatemala (Thillet et al, 2003).

Poverty limits local peoples' resource management alternatives

Protecting the forest cannot simply mean keeping local people out. But simple solutions rarely exist. When limits need to be placed on forest use, there must be alternative sources of income for forest-dependent people (Bazaara, 2003).

Outcomes of current policies

Increased local authority over forests can improve stability or hasten deforestation

In India, in Kumaon and other areas where local elected *van panchayats* manage forests, forests are in very good condition and have provided local livelihoods for 60 years (Agrawal, 2002; Sarin et al, 2003). At the other extreme, Indonesian forests under greater local control have seen a proliferation of logging contracts with little concern for environmental effects (Resosudarmo, 2004). In Yunnan, China, where local authorities were given new powers over forest resources during the past, there was also a dramatic increase in deforestation (Dachang and Edmunds, 2003). But this increase turned out to be temporary, and the trend reversed. Analysis suggests that the increase was due to tenure insecurity and the fear that the rights to manage forests would again be taken away (Dachang and Edmunds, 2003) – fears that are also prevalent in Indonesia.

Bazaara (2003) concludes that environmental outcomes of decentralization depend upon whether it is in the interest of dominant groups to protect the environment and the existence of effective checks and balances.

Decentralization policies have positive social effects when they empower local people

In Bolivia, local people now have legal and secure access to forest resources that they did not have previously; local governments also have more resources through both government transfers and the redistribution of forestry funds; and local groups with access to forest resources have greater negotiating capacity with absent landowners and concessionaires (Pacheco, 2003). Indigenous and peasant interests are taken into account where these actors participate in local government (Larson, 2003b; Pacheco, 2003) or where personal or political ideologies are based upon support to these sectors (Toni, 2003). Incomes are high and forests are well managed in China where village committees have administered resources accountably and responsibly (Dachang and Edmunds, 2003). Creative solutions have provided positive social outcomes in Uganda, such as organizing locals to manage resource exploitation and improve their incomes, then taxing them to benefit the municipality as a whole (Bazaara, 2003).

Even in Indonesia, where increased local authority has clearly had detrimental effects on forests, local people have significant income from forests for the first time in 30 years. At the same time, this has mostly been captured by those who had capital (Resosudarmo, 2004). Nevertheless, even local entrepreneurs, unlike outside logging companies in the past, are under social pressure to contribute to local projects (McCarthy, 2001b).

Decentralization policies have negative social effects when they extend state control over local people

Where state policies undermined local institutions, women, ethnic minorities and the very poor have not benefited from devolution policies. In India, in places where the state had previously imposed tight restrictions, subsistence access to forests did, in fact, increase; but where there were local management institutions previously, state contracts decreased local access (Edmunds et al, 2003).

In India and the Philippines, the forest department determined the species for reforestation and afforestation, displacing species valued locally. Plantations were established on common lands that people had used for other purposes for livelihood needs (Edmunds et al, 2003). Local governments did the same thing in China where they were not accountable to villagers (Dachang and Edmunds, 2003). Decentralization that ignores existing institutions creates income opportunities for the elite but destroys them for the poor (Edmunds et al, 2003).

Decentralization may actually strengthen the standing of the elite (McCarthy, 2001a; Pacheco, 2003). In some cases the elite apply the law against those who do not have the political contacts or capital to legalize their activities (McCarthy; 2001a). The central government may also form alliances with local elites to maintain control (Edmunds et al, 2003).

Failure to address equity issues can lead to new conflicts. Small-scale permits in Indonesia may generate conflicts between local illegal loggers and the elite (Resosudarmo and Dermawan, 2002). Special provisions are needed to address these inequities and to enable equal access to decision-making (Dupar and Badenoch, 2002). Disadvantaged groups require provisions to ensure that their livelihood interests and rights are protected (Sarin et al,

2003). Management objectives should be designed with broad-based participation (Sarin et al, 2003). Some researchers argue that direct devolution to local resource users, rather than through local governments, is more effective (Shackleton et al, 2002).

In some cases, decentralization policies with negative effects on local people had positive effects on forest cover. Most researchers believe that increasing regeneration while decreasing access for forest-dependent poor people is not an acceptable alternative, especially if trees are still felled for state revenue (Contreras, 2003; Edmunds et al, 2003; Sarin et al, 2003).

DEMOCRATIC DECENTRALIZATION THAT WORKS

We conclude by summarizing the core issues that have arisen from the case studies and make suggestions for the future.

First, the case studies reveal a remarkable lack of information about the legal framework for forestry or for decentralization. In many cases, not even local governments understand their rights and responsibilities. This makes it very difficult for local governments to fulfil their duties and for citizens to hold them accountable.

Second, the environment and natural resources comprise a highly conflictive arena when it comes to the distribution of power, as well as a domain that is particularly vulnerable to corruption. Decentralization protagonists must take political and economic realities into account.

Third, checks and balances are needed on all levels of power and authority. Elites benefit most often whether forest management is centralized or decentralized. Marginalized groups tend to lose out unless they receive special attention and are specifically included.

Finally, the monitoring of logging and management plans and the enforcement of forest management regulations are generally rare or highly selective occurrences. Few authorities – central or local – appear willing or able to prosecute loggers, particularly logging companies, who violate forest rules. Enforcement is sometimes directed against those who are easier to control rather than those who commit the most serious violations.

The case studies also reveal that democratic decentralization is hard to come by. If democratic decentralization requires significant discretionary powers in the hands of downwardly accountable local actors, then current processes often fail on both counts. There are several fundamental problems. One is the resistance on the part of central governments to give up powers, or the tendency to give them to local entities which it can control. Another is the weakness of current electoral processes, as well as other downward accountability mechanisms, to guarantee the representativeness, transparency and accountability of local governments. A third is the failure to recognize decentralization as a value-laden process for good governance and resources management in the common interest (Oyono, 2004), rather than simply a way of cutting costs.

A central tenet of democratic decentralization is the importance of taking into account local social, ecological and cultural dynamics – adjusting national policy to the particularities of each district. Rather than defining a blanket mechanism for forest management, decentralized administration should make it possible to identify and respond to local differences. What is the combination of incentives that coalesce in each local arena towards sustainable or unsustainable forest management? Forestry decentralizations should begin by working with institutions that are already there – working *with* local people, and building on what they have already built. Elected, representative and accountable local authorities are the most likely interlocutors for this 'localization' process.

The lessons discussed above suggest alternatives for building a democratic decentralization that works for local people and forests:

- The decentralization experiment needs to be implemented: meaningful discretionary powers over forests should be turned over to representative and accountable local institutions under a clear and unified legal framework.
- Electoral processes should allow for independent local candidates, and multiple accountability mechanisms should be promoted and strengthened.
- Effective legal recourse mechanisms should be established to promote government accountability at all levels.
- Representative and effective institutions should be supported to build greater consensus among stakeholders at all levels, with particular emphasis on promoting the participation of marginalized groups.
- Information regarding logging contracts and funding should be managed transparently at all levels, and local governments and local people should benefit from income generated by forest resources.
- Central governments should be effective partners in decentralization by training local governments and local peoples in both forest management and local democratic processes. They should also provide oversight of local officials.
- Forestry should be recognized as a multifaceted sphere that is not just economic but also ecological and social, and forestry professionals should be trained in all of these facets.
- Third parties such as NGOs should help to promote decentralization by raising the voice of local people in defence of their interests.
- Elected local governments should be encouraged to build regional associations to address broader-scale issues and problems.

These 'solutions', however, do not stand outside of the existing political economy of each nation. If they did, they would probably already have been implemented much more fully. Hence, efforts to decentralize must recognize the social, economic and political interests of each set of actors and begin to build a political climate that makes real decentralization possible. Broad coalitions of local actors, NGOs, donors, local governments and sympathetic central government officials are needed to counteract the centralizing tendencies of central governments (Ribot et al, 2005). And local actors must become effective players in their own right – in order to demand decentralization and that it should be implemented in their interest. Given the experiences of the case studies reviewed here, it appears likely that without significant pressure from their citizens, central governments may continue to fail giving priority to the 'democratic' in decentralization.

Note

This chapter would not have been possible without being able to draw upon the excellent work of numerous researchers and other professionals from around the world, as well as upon conversations with many of them. I particularly want to thank Arun Agrawal, Yovanni Alvarado, Mynor Barillas, Carol Colfer, Lyes Ferroukhi, David Kaimowitz, Lungisile Ntsebeza, René Oyono, Pablo Pacheco, Daju Resosudarmo, Jesse Ribot, Mario Vallejo and Lini Wollenberg, as well as all of the participants of the Bellagio Workshop on Decentralization and the Environment. I also want to thank the Center for International Forestry Research (CIFOR), the Swedish International Development Cooperation Agency (Sida) – International Development Bank (IDB) Partnership, the Ford Foundation and Nitlapán-Central American University (UCA) for supporting my research over the past few years. Two anonymous reviewers gave helpful comments on an earlier draft.

References

Agrawal, A. (2001) 'The decentralizing state: Nature and origins of changing environmental policies in Africa and Latin America 1980–2000'. Paper prepared for the 97th Annual Meeting of the American Political Science Association, 30 August–2 September, San Francisco, California

Agrawal, A. (2002) 'The regulatory community: Decentralization and the environment in the *van panchayats* (forest councils) of Kumaon'. Paper presented at the World Resources Institute Workshop on Decentralization and the Environment, 18–22 February, Bellagio, Italy

Agarwal, A. and Ribot J. (1999) 'Accountability in decentralization: A framework with South Asian and West African environmental cases'. *The Journal of Developing Areas*, vol 33, pp473–502

Barr, C., Wollenberg, E., Limberg, G., Anau, N., Iwan, R., Sudana, I. M., Moeliono, M. and Djogo, T. (2001) *The Impacts of Decentralizations on Forests and Forest-dependent Communities in Malinau District, East Kalimantan*. CIFOR, Bogor, Indonesia

Baviskar, A. (2004) 'Between micro-politics and administrative imperatives: Decentralization and the watershed mission in Madhya Pradesh, India', in Ribot, J. and Larson, A. (eds) *Democratic Decentralization through a Natural Resource Lens, Special Issue, European Journal of Development Research*, vol 16, no 1, pp26–40

Bazaara, N. (2003) 'Decentralization, politics and environment in Uganda', *Environmental Governance in Africa Working Paper Series No 7*, World Resources

Institute, Washington, DC
Beneria-Surkin, J. (2003) *Decentralization Questioned: The Structuring and Articulation of Guarani Participation in Conservation and Development in Izozog, Bolivia*. PhD thesis, University of California, Los Angeles
Bigombé Logo, P. (2003) 'The decentralized forestry taxation system in Cameroon: Local management and state logic', *Environmental Governance in Africa Working Paper Series No 10*, World Resources Institute, Washington, DC
Carney, D. and Farrington, J. (1998) *Natural Resource Management and Institutional Change*. Routledge, London
Chambers, R. (1997) *Whose Reality Counts? Putting the First Last*. ITDG Publishing, London
Colfer, C. (2005) *The Complex Forest: Community, Uncertainty and Adaptive Collaborative Management*. Resources for the Future and CIFOR, Washington, DC, forthcoming
Contreras, A. (2003) 'Creating space for local forest management: The case of the Philippines', in Edmunds, D. and Wollenburg, E. (eds) *Local Forest Management: The Impacts of Devolution Policies*. Earthscan, London, pp127–149
Cousins, B. and Kepe, T. (2004) 'Decentralization when land and resource rights are deeply contested: A case study of the Mkambati Eco-tourism project on the wild coast of South Africa', in Ribot. J. and Larson, A. (eds) *Democratic Decentralization through a Natural Resource Lens, Special Issue, European Journal of Development Research*, vol 16, no 1, pp41–54
Crook, R. and Manor, J. (1998) *Democracy and Decentralisation in South Asia and West Africa*. Cambridge University Press, Cambridge
Crook, R. C. and Sverrisson, A. S. (2001) *Decentralization and Poverty Alleviation in Developing Countries: A Comparative Analysis or, is West Bengal Unique?* IDS Working Paper 130, Institute for Development Studies, Brighton
Dachang, L. and Edmunds, D. (2003) 'The promises and limitations of devolution and local forest management in China', in Edmunds, D. and Wollenburg, E. (eds) *Local Forest Management: The Impacts of Devolution Policies*. Earthscan, London, pp20–54
deGrassi, A. (2003) *Constructing Subsidiarity, Consolidating Hegemony: Political Economy and Ago-Ecological Processes in Ghanaian Forestry*. Environmental Governance in Africa Working Paper No 13, World Resources Institute, Washington, DC
Dupar, M. and Badenoch, N. (2002) *Environment, Livelihoods and Local Institutions*. World Resources Institute, Washington, DC
Edmunds, D. and Wollenburg, E. (eds) (2003) *Local Forest Management: The Impacts of Devolution Policies*. Earthscan, London
Edmunds, D., Wollenburg, E., Contreras, A., Dachang, L., Kelkar, G., Nathan, D., Sarin, M. and Singh, N. (2003) 'Introduction', in Edmunds, D. and Wollenburg, E. (eds) *Local Forest Management: The Impacts of Devolution Policies*. Earthscan, London, pp1–19
Ellis, F. (2000) *Rural Livelihoods and Diversity in Developing Countries*. Oxford University Press, Oxford
Enters, T. and Anderson, J. (1999) 'Rethinking the decentralization and devolution of biodiversity conservation'. *Unasylva*, vol 50, no 4, www.fao.org/documents/show_cdr.asp?url_file=/docrep/x3030e/x3030e00.htm
Ferroukhi, L. (ed) (2003) *Gestión Forestal Municipal en América Latina*. CIFOR/IDRC, Bogor, Indonesia

Ferroukhi, L. and Aguilar, A. (2003) 'Avances y Desafíos de la Gestión Forestal Municipal en Costa Rica', in Ferroukhi, L. (ed) *Gestión Forestal Municipal en América Latina*. CIFOR/IDRC, Bogor, Indonesia, pp179–212

Ferroukhi, L. and Echeverría, R. (2003) 'Las Políticas de Gestión Forestal Descentralizada en Guatemala', in Ferroukhi, L. (ed) *Gestión Forestal Municipal en América Latina*. CIFOR/IDRC, Bogor, Indonesia, pp89–112

Gibson, C., McKean, M. and Ostrom, E. (2000) 'Explaining deforestation: The role of local institutions', in Gibson, C., McKean, M. and Ostrom, E. (eds) *People and Forests: Communities, Institutions, and Governance*. MIT Press, Cambridge, MA

Kaimowitz, D. (2000) *El Papel de las Corporaciones Municipales en las Actividades de Recursos Naturales del Proyecto de Administración de Areas Rurales*. Report, Rural Area Administration Project (PAAR), Tegucigalpa

Kaimowitz, D. and Ribot, J. (2002) 'Services and infrastructure versus natural resource management: Building a base for democratic decentralization'. Paper presented at the World Resources Institute Workshop on Decentralization and the Environment, Bellagio, Italy, 18–22 February

Kassibo, B. (2002) 'Participatory management and democratic decentralization management of the Samori Forest in Babye commune, Mopti region, Mali'. Paper presented at the World Resources Institute Workshop on Decentralization and the Environment, Bellagio, Italy, 18–22 February

Kassibo, B. (2003) *Historical and Political Foundations for Participative Management and Democratic Decentralization in Mali: A Synthesis of Two Case Studies*. World Resources Institute, Washington, DC

Klooster, D. (2002) 'Towards adaptive community forest management: Integrating local forest knowledge with scientific forestry'. *Economic Geography*, vol 78, no 1, pp43–70

Larson, A. (2001a) *Rainforest Conservation and Grassroots Development: If Ever the Twain Shall Meet? Peasant Colonists and Forest Conversion in the Nicaraguan Rainforest*. PhD thesis, University of California, Berkeley

Larson, A. (2001b) *Recursos Forestales y Gobiernos Municipales en Nicaragua: Hacia una Gestión Efectiva*. CIFOR/Nitlapan, Managua, Nicaragua

Larson, A. (2002) 'Natural resources and decentralization in Nicaragua: Are local governments up to the job?' *World Development*, vol 30, no 1, pp17–31

Larson, A. (2003a) 'Decentralization and forest management in Latin America: Toward a working model'. *Public Administration and Development*, vol 23, no 2, pp211–226

Larson, A. (2003b) 'Gestión Forestal Municipal en Nicaragua: Descentralización de Cargas, Centralización de Beneficios?', in Ferroukhi, L. (ed) *Gestión Forestal Municipal en América Latina*. CIFOR/IDRC, Bogor, Indonesia, pp113–144

Larson, A. (2003c) 'Local governments and forest management in Central America: Conceptions of decentralization in national context'. Paper presented at the Open Meeting of the Human Dimensions of Global Environmental Change Research Community, International Science Planning Committee, Montreal, Canada, 16–18 October 2003

Larson, A. (2003d) *Mejores Prácticas para la Gestión Descentralizada de los Recursos Forestales en Centroamérica*. Report for the SIDA-IDB Partnership, 20 June, Managua

Larson, A. (2004) 'Formal decentralization and the imperative of decentralization "from below": A case study of natural resource management in Nicaragua', in Ribot, J. and Larson, A. (eds) *Democratic Decentralization through a Natural*

Resource Lens, Special Issue, *European Journal of Development Research*, vol 16, no 1, pp55–70

Larson, A. (2005) Indigenous Communities and Local Representation in Forestry Decentralizations: Case Studies from Nicaragua and Guatemala. Research report, WRI/CIFOR, Washington, DC, forthcoming

Larson, A. and Ferroukhi, L. (2003) 'Conclusiones', in Ferroukhi, L. (ed) *Gestión Forestal Municipal en América Latina.* CIFOR/IDRC, Bogor, Indonesia, pp213–233

Lutz, E. and Caldecott, J. (1996) 'Introduction', in Lutz, E. and Caldecott, J. *Decentralization and Biodiversity Conservation*, World Bank symposium, World Bank, Washington, DC

Magno, F. (2001) 'Forest devolution and social capital: State–civil society relations in the Philippines'. *Environmental History*, vol 6, no 2, pp264–286

Mapedza, E. and Mandondo, A. (2002) *Co-management in the Mafungautsi State Forest Area of Zimbabwe: What Stake for Local Communities?* Environmental Governance in Africa Working Paper No 5, World Resources Institute, Washington, DC

Mandondo, A. (2000) *Situating Zimbabwe's Natural Resource Systems in History.* Occasional paper no 32, CIFOR, Bogor, Indonesia

Mandondo, A. and Mapedza, E. (2003) *Allocation of Governmental Authority and Responsibility in Tiered Governance Systems: The Case of Environment-Related Laws in Zimbabwe.* Environmental Governance in Africa Working Paper No 11, World Resources Institute, Washington, DC

Manor, J. (1999) *The Political Economy of Democratic Decentralization.* World Bank, Washington, DC

Manor, J. (2004) 'User committees: A potentially damaging second wave of decentralization?' in Ribot, J. and Larson, A. (eds) *Democratic Decentralization through a Natural Resource Lens, Special Issue*, *European Journal of Development Research*, vol 16, no 1, pp192–213

McCarthy, J. (2001a) *Decentralization and Forest Management in Kapuas District, Central Kalimantan,* Case Study 2. CIFOR, Bogor, Indonesia

McCarthy, J. (2001b) *Decentralization, Local Communities and Forest Management in Barito Selatan District, Central Kalimantan,* Case Study 1. CIFOR, Bogor, Indonesia

Meynen, W. and Doornbos, M. (2004) 'Decentralizing natural resource management: A recipe for sustainability and equity?' in Ribot, J. and Larson, A. (eds) *Democratic Decentralization through a Natural Resource Lens, Special Issue*, *European Journal of Development Research*, vol 16, no 1, pp235–254

Muhereza, F. (2003) *Commerce, Kings and Local Government in Uganda: Decentralizing Natural Resources to Consolidate the Central State.* Environmental Governance in Africa Working Paper Series No 8, World Resources Institute, Washington, DC

Ntsebeza, L. (2004) 'Democratic decentralization and traditional authority: Dilemmas of land administration in rural Africa', in Ribot, J. and Larson, A. (eds) *Democratic Decentralization through a Natural Resource Lens, Special Issue*, *European Journal of Development Research*, vol 16, no 1, pp71–89

Oyono, R. (2003) *Institutional Deficit, Representation and Decentralized Forest Management in Cameroon: Elements of Natural Resources Sociology for Social Theory and Public Policy*, Environmental Governance in Africa Working Paper Series No 15. World Resources Institute, Washington, DC

Oyono, R. (2004) 'Social and organizational roots of ecological uncertainties in Cameroon's forest management decentralization model', in Ribot, J. and Larson, A.

(eds) *Democratic Decentralization through a Natural Resource Lens, Special Issue, European Journal of Development Research*, vol 16, no 1, pp174–191

Pacheco, P. (2003) 'Municipalidades y Participación Local en la Gestión Forestal en Bolivia', in Ferroukhi, L. (ed) *Gestión Forestal Municipal en América Latina.* CIFOR/IDRC, Bogor, Indonesia, pp19–56

Pacheco, P. (2004) 'What lies behind decentralization? Forest, powers and actors in lowland Bolivia', in Ribot, J. and Larson, A. (eds) *Democratic Decentralization through a Natural Resource Lens, Special Issue, European Journal of Development Research*, vol 16, no 1, pp90–109

Peluso, N. (2002) 'Some questions about violence and decentralization: A preliminary exploration of the case of Indonesia'. Paper presented at the World Resources Institute Workshop on Decentralization and the Environment, Bellagio, Italy, 18–22 February

Pommier, D. (undated) 'Barreras a la Legalidad, Buen Gobierno del Sector Forestal e Impacto Sobre la Pobreza en Nicaragua', available at www.talailegal-centroamerica.org/downloads/barreras_nicaragua_sp6_lowres.pdf (19 November 2003)

Resosudarmo, I. A. P. (2004) 'Closer to people and trees: Will decentralization work for the people and forests of Indonesia?' in Ribot, J. and Larson, A. (eds) *Democratic Decentralization through a Natural Resource Lens, Special Issue, European Journal of Development Research*, vol 16, no 1, pp110–132

Resosudarmo, I. A. P. and Dermawan, A. (2002) 'Forests and regional autonomy: The challenge of sharing the profits and pains', in Colfer, C. and Resosudarmo, I. A. P. (eds) *Which Way Forward? People, Forests and Policymaking in Indonesia.* Resources for the Future and CIFOR, Washington, DC

Ribot, J. (2002) *Democratic Decentralization of Natural Resources: Institutionalizing Popular Participation.* World Resources Institute, Washington, DC

Ribot, J., Agrawal, A. and Larson, A. (2005) 'Resisting decentralization of natural resource management: New institutionalism meets political economy'. *World Development*, forthcoming

Romeo, L. (2003) 'The road from decentralization to poverty reduction and the role of external assistance'. *Public Administration and Development*, vol 23, no 1, pp89–96

Sarin, M., Singh, N., Sundar, N. and Bhogal, R. (2003) 'Devolution as a threat to democratic decision-making in forestry? Findings from three states in India', in Edmunds, D. and Wollenburg, E. (eds) *Local Forest Management: The Impacts of Devolution Policies.* Earthscan, London, pp55–126

Shackleton, S., Campbell, B., Wollenberg, E. and Edmunds, D. (2002) 'Devolution and community-based natural resource management: Creating space for local people to participate and benefit?' *ODI Natural Resource Perspectives*, vol 76, March

Sithole, B. (2004) 'Becoming men in our dresses! Women's involvement in a joint forestry management project in Zimbabwe', in Colfer, C. (ed) *The Equitable Forest: Diversity, Community and Resource Management.* Resources for the Future and CIFOR, Washington, DC

Thillet, B. et al (2003) *Tierras Municipales en Guatemala: Un Desafío para el Desarrollo Local Sostenible.* FLACSO, Guatemala

Toni, F. (2003) 'La Gestión Forestal en los Municipios de la Amazonia Brasileña', in Ferroukhi, L. (ed) *Gestión Forestal Municipal en América Latina.* CIFOR/IDRC, Bogor, Indonesia, pp145–178

Vallejo, M. (2003) 'Gestión Forestal Municipal: Una Nueva Alternativa para Honduras', in Ferroukhi, L. (ed) *Gestión Forestal Municipal en América Latina*, CIFOR/IDRC, Bogor, Indonesia, pp57–88

Vallejo, M. and Ferroukhi L. (2005) *Municipios y Bosques: Alternativa para el Desarrollo*. Ediciones El SEDAL y Guardabarranco,Tegucigalpa

Wittman, H. (2002) *Negotiating Locality: Decentralization and Communal Forest Management in the Guatemalan Highlands*. Masters thesis, Cornell University, Ithaca, NY

World Bank (1988) *World Development Report*. Oxford University Press, New York

World Bank (1997) *World Development Report*. Oxford University Press, New York

World Bank (2000) *World Development Report 1999/2000: Entering the 21st Century*. Oxford University Press, New York

Xu, J. and Ribot, J. (2004) 'Decentralization and accountability in forest management: Case from Yunnan, Southwest China', in Ribot, J. and Larson, A. (eds) *Democratic Decentralization through a Natural Resource Lens, Special Issue, European Journal of Development Research*, vol 16, no 1, pp153–173

Chapter 3

Paths and Pitfalls of Decentralization for Sustainable Forest Management: Experiences of the Asia Pacific Region

Ian Ferguson and Cherukat Chandrasekharan

INTRODUCTION

Decentralization is widely mooted as a device to improve the democratic process of governance and the equity and/or efficiency of resource allocation. While the objectives and the key principles of decentralization, such as subsidiarity, are well known, attempts to implement them have not always been successful. Selected case studies of experiences in the Asia Pacific region, reviewed for their positive and negative aspects, suggest that the paths for successful decentralization need to reflect the geography, culture and institutions of the country concerned, as well as the various functions of governance, bureaucracy, markets and information and their interrelationships. Decentralization can embrace geographic hierarchies, as well as joint federal–regional, inter-regional or local government collectives, and other arrangements that do not conform to normal geographic hierarchies of governance. However, the preparatory demands are high in all forms and suggest that phasing implementation is desirable. The lessons learned from the case studies provide a basis for charting future paths and avoiding pitfalls in implementing decentralization for sustainable forest management.

THE IMPETUS FOR DECENTRALIZATION

Much of the current impetus for decentralization stems from concerns regarding democratic governance. Summing up the perceived advantages of decentralization, Furtado (2001) wrote that it would:

- ensure the provision of social services that meet the needs of constituents in a given locale;
- draw on local knowledge and preferences;

- give interests at local levels a stronger sense of ownership over projects and programming, thus making programming more sustainable;
- enhance the public accountability of bureaucrats, elected representatives and political institutions, thus ensuring greater responsiveness in government;
- promote local self-reliance; and
- promote monitoring, evaluation and planning at the local level and enhance community participation in decision-making.

The idea also had its detractors. Critics feared that the decentralization of public services would result in an erosion of quality and consistency across regions, leaving some regions, cities, villages and potentially vulnerable groups worse off than others.

Today, there is a strong push in favour of decentralization to promote improved governance or to combat corruption. However, there is no evidence that greater decentralization automatically leads to these results. The democratic process itself is a work in progress, not a known and fixed target. There is, therefore, no one path, and decentralization can take many forms. For simplicity, let us recognize three forms (Klugman, 1994) and one variant:

1 *Deconcentration*: the transfer of administrative responsibility for specified functions to lower levels within the central government bureaucracy, generally on some spatial basis.
2 *Delegation*: the transfer of managerial responsibility for specified functions to other public organizations outside normal central government control, whether provincial or local government or parastatal agencies.
3 *Devolution*: the transfer of governance responsibility for specified functions to sub-national levels, either publicly or privately owned, that are largely outside the direct control of the central government.
4 *Privatization*: a particular form of devolution to private ownership.

In order to avoid confusion in the terminology of the sub-national geographic basis of decentralization, we have adopted the following generic hierarchy:

- *provincial*: first-order geographic subdivision of the national level;
- *district*: second-order geographic subdivision of the provincial levels;
- *village*: third-order geographic subdivision of the district levels;
- *household*: independent livelihood units forming a village.

In this review of case studies from countries in the Asia Pacific region, reference will be made to *unitary* and *federal* systems of governance. In a unitary system, the central government defines the powers of other levels of government. In a federal system, the powers of local governments derive from the constitution and are not subject to change by central government (see Chapter 1). Federal systems are often established to create national unity without eliminating traditional local autonomy (for example, Switzerland). Unitary systems also have had sub-national levels of governance; but these are not constitutionally empowered – they are instituted to balance the burden of governance.

Decentralization to *community-based units* (groups of households in close proximity to forestland) needs to be recognized as a further category. The fundamental difference between this and decentralization to a district or village level is that these communities are not necessarily coincident with either administrative district or village boundaries. Community-based units include a variety of non-governmental membership organizations such as co-operatives, farmers' associations and tenant leagues; based on voluntary participation and group solidarity, they are generally non-bureaucratic and are not substitutes for public or private-sector channels of service, investment or activity.

Decentralization through *customary ownership* of land represents another special case. Customary or communal ownership, in which every individual or household in the community has a right over the forest property, is a decentralized situation. Customary ownership is the common form of land tenure in the Pacific Island countries.

Countries in the Asia Pacific region have taken diverse paths to decentralization, and the institutions and property rights involved vary widely. The change in the forest area from 1990 to 2000 is negative* in 13 of 20 countries – Australia;* Bangladesh; Cambodia;* China; Fiji;* India; Indonesia;* Republic of Korea;* Lao People's Democratic Republic;* Malaysia;* Mongolia;* Myanmar;* Nepal;* New Zealand; Papua New Guinea; the Philippines;* Sri Lanka;* Thailand;* Vietnam; and Vanuatu – and even some of these may show improvement only because of changes in definition, such as the inclusion of rubberwood plantations. Thus, the changes in forest area are not indicative of major improvement in one of the simplest (albeit crude) indicators of sustainable forest management.

The following case studies have been chosen to illustrate some of the paths that have been pursued to achieve decentralization and some of the pitfalls associated with those changes.

DEVOLUTION TO DISTRICT GOVERNANCE

The Philippines

In the Philippines there have been several initiatives since the 1986 election with regard to democratic decentralization and devolution of authority to a hierarchy of units, together with special arrangements to link community-based forestry projects to those units. The intention of the reorganization was for the national functions of the Department of Environment and Natural Resources to be decentralized to the regional, provincial and district levels, and that environmental and natural resources issues should be addressed in an integrated and holistic manner. In the process, the national agency was to be transformed from a regulatory and controlling organization to a mainly extension organization, with greatly reduced regulatory functions. The goal, however, is far from being achieved.

The Department of Environment and Natural Resources is a huge, decentralized and (partially) national organization with a bureaucratic style of communications. Decisions at the top and implementation on the ground are often separated by considerable time lags; feedback about field activities reaches headquarters only after long delays, impeding quick managerial decisions; and monitoring and evaluation of projects are extremely difficult.

The Local Government Code granted the new hierarchy of governance units greater fiscal and political autonomy, thereby expanding their capacity to participate in national development efforts. It has brought to the fore the critical role of district units in managing forest and watersheds. In general, the code has tasked them to adopt measures that will 'protect the environment and impose appropriate penalties for acts which endanger the environment'. More specifically, the code gives responsibility to the appropriate levels of units in the governance hierarchy, as follows:

- village (*barangay*) level: services and facilities related to general hygiene and sanitation, beautification and solid waste collection;
- district (municipality) level: implementation of community-based forestry projects, management and control of communal forests not exceeding 50 square kilometres and the establishment of tree parks and similar forest projects; and
- provincial level: the enforcement of forestry laws limited to community-based forestry projects.

Other legislation has transferred certain forestry development functions to districts for implementation. The closeness of districts to the local people and their proximity to the forest resources make them powerful partners of the national forestry agency in implementing community-based forest management (SUSTEC, 2001).

The national forestry agency has already devolved some of its conservation, management and protection functions to districts and reassigned some 1000 staff members to support them. The decentralization did not result in the proper integration of field activities and staff (Chandrasekharan, 2003). It was also planned that the agency would transfer budgets, assets, and records that correspond to the department's devolved functions and programmes; but there has been only limited progress in this. Many district units are attempting to defer the devolution, citing a lack of clarity in the new responsibilities of local authorities and of funding, inadequate office space to accommodate new staff, and complexities in administrative arrangements.

District functionaries believe that they should have power to make important decisions on resources management, but real power (they say) remains at the centre. In most districts there is no forestry expertise or capability to deal with forestry matters. Forest-related issues often impinge on the life of the local people and their livelihood activities; but districts are, in most cases, unable to intervene. Substantial differences exist in the policy thrusts, priorities and approaches of the national forestry agency and districts relating to the management and utilization of forests, leading to weaknesses in collaboration.

Community-based organizations have similar problems, as the case study in Chapter 16 illustrates. In the Philippines, all people-oriented forestry programmes were integrated within the community-based forest management programme in 1995. The programme involves a partnership between the national government and the forest communities in developing, rehabilitating and managing vast tracts of forest areas. It recognizes that food and sustainable livelihoods, rather than forestry, are the chief priorities of upland dwellers. By giving community-based organizations responsibilities for resource management, community-based forest management seeks to mitigate inappropriate farming techniques, shifting cultivation and forest exploitation; to promote sustainable management of forest resources, social justice and improved well-being of local communities; and to develop strong partnership among local communities and the national forestry agency.

Community-based forest management applies to all areas classified as forestlands, including allowable zones within protected areas not covered by prior vested rights. As of 2002, some 4956 sites were covered by these schemes, for a total of 5.7 million hectares (ha). The programme has benefited 2182 project areas constituting 496,175 households with about 2.5 million community members.

The community-based forest management committees and the national federation of those organizations are generally, as yet, ineffective paper bodies (Chandrasekharan, 2003), notwithstanding the qualified success of some (see Chapter 16). The problem may lie not with the policy itself, but in the administrative institutions: senior staff in bureaucratic structures appear reluctant to surrender decision-making power.

Indonesia

Rapid devolution of powers from a national government to district governments (*kabupaten*) has taken place in Indonesia (see Chapters 1 and 7). The laws stipulate that district government will be the main functional level of decentralized government. Only national parks and nature reserves are to be managed directly by the national government; all other forestry activities are to be under the district government. Currently, however, the district governments are unprepared to manage forestry (Suwondo, 2002).

Dauvergne (1997) and Brown (1999) have documented extraordinary levels of corruption: at least 40 per cent and perhaps as much as 70 per cent of timber was cut illegally (Scotland and Ludwig, cited by Brack et al, 2002).

Giving district governors the powers to allocate concessions to local cooperatives might create a new system of corruption at the local governance level; in fact, according to Suwondo (2002), corruption has often shifted to district government. Suwondo also notes the case of Lombok, where local governance took over a logging concession to better effect than before.

The rapid devolution of responsibility for allocating concessions, coupled with other shifts in national government policies related to forests, has created confusion in the sector. Rukmantara (2003), for example, reported that 'the decentralization of the process by which small-scale operators were granted concessions caused conflicts on the ground because such concessions often overlapped with those being managed by existing forest concessions'. A short-term ban on legal logging would allow authorities to identify the offenders and institute controls; but this is only likely if public concern is raised sufficiently during national elections.

Decentralization to village governance

Korea

Since 1908, under the Great Korean Empire, forestry administration has been under the Ministry of Agriculture, Forestry and Fisheries. Forestry administration follows a de-concentrated internal structure; it executes laws relating to forests and forestry, manages forest resources and the utilization of forest products, conducts research and training and runs an extension service.

At its constitutional formation in 1948, the forest landscape of the Republic of Korea was badly degraded and denuded. Some 73 per cent of the forest area of the Republic of Korea is under private ownership; the average size of holdings is 2.6ha. Small plots are managed as part of the system of village co-operatives (village forestry associations), with the strong support of the government. Thus, forestry development in Korea involves village-level cooperation in achieving community goals. The system consists of a hierarchy of forestry associations at the provincial and national levels involved in all production and technical aspects of forestry, with higher tiers having supervisory and technical guidance functions. The system is often cited as an example of a successful scheme of a co-operative, multi-tiered and federated structure, with a comprehensive programme involving social mobilization and covering all aspects of forestry for the improvement of community life.

In 1973 Korea embarked on an ambitious programme of national forestry development, which was carried out by the village cooperatives. It combined voluntary decision-making by village co-operatives with a national programme for forestry development and a federation imposed from above (FAO, 1982). The fundamental philosophy involves a set of procedures: discussion of the problem, election of leaders, formulation of plans in the village assembly, and undertaking of projects with grassroots cooperation through voluntary participation without discrimination on gender or age. As of 1990 there were 20,287 village co-operatives grouped within 142 provincial-level units, whose membership reached 1,983,600. The multi-tiered system involved significant pieces of forestry and land-use legislation, which taken as a package provided national guidance and directives for a comprehensive programme. It involved government at all levels through provisions of subsidies, loans and technical support, and cooperation with private organizations – thus linking planning and action between government and village-level organizations through clear lines of authority and interaction. During the 1980s and 1990s, the government supported about 770 technical forest guides at the provincial forestry association level to help forest owners and to provide them with advanced technology.

Forest owners and villages subsequently strengthened the system by forming self-propelling co-operatives of more homogenous membership. The change was seen to improve efficiency through systematically improved use of forests, establishment of forest labour units, enlargement of the forest fund and the creation of a stronger forest sector driven by private initiatives (Yoo, 1997). The system now has only two higher coordinating levels: the unit level in 142 provinces and the national level. Even though the number of cooperatives is lower, the development potential is greater because the members participate more actively. The ratio of extension visits by extension agents increased about five times in two years after the transformation, and the national government has increased financial support.

Nepal

After eliminating the feudal tenure system in 1957 through the Private Forest Nationalization Act, which brought forests under public domain, Nepal experimented with systems to decentralize forestry activities to village or community level (generally with a population of 2000 to 4000) through different routes. The Forest Act of 1961 provided recognition of village (*panchayat*) forests. The Forest Preservation Act of 1967 recognized village-protected forests. The Forest Policy of Nepal of 1978 allowed national forests to be handed over to the care of villages. The villages were made responsible for planting and protection of trees in village forests (which are government-owned wastelands) and in return obtained all rights to the produce. Villages also managed the protected forests following a viable forest management system and in return collected fuelwood, fodder and non-wood forest products, receiving 75 per cent of any revenue.

During the early 1980s, Nepal passed decentralization legislation (the Local Government Act) empowering village government units to make many decisions previously decided by the national government. Village units were linked to the district government units that had to submit budgets and programmes to a provincial-level unit for approval.

During the late 1980s the national government recognized forest communities (forest user groups), irrespective of village political boundaries, as more appropriate institutions for sustainable forest management. The Forest Act of 1993 and the forest regulations framed in 1995 reaffirmed the national government's policy in assigning more responsibility to the communities by recognizing them as self-governing institutions with rights to acquire, sell and transfer forest products (Singh and Kafle, 2000).

Meeting the community's basic needs is the priority, and all the benefits from managing the forests go to the communities. Their responsibilities are to prepare annual plans to manage forest; protect, manage and utilize forest produce; fund their own activities and receive grants; reinvest at least 25 per cent of revenues in forest development; and set punishments for members who violate rules.

Decision-making by village government units is by consensus, but often excludes women and marginalized ethnic and caste groups (Dangol, 2005). By March 2000, approximately 9000 communities had legally taken over management responsibilities for about 12 per cent of Nepal's forestlands. The community units have established a federation of community forestry users to expand and strengthen their role in managing Nepal's forests. At present the federation's activities include training, networking, advocacy and information dissemination. The prime responsibility for extension delivery and enhancing outreach capability lies with the district staff of the national forestry agency.

Decentralization in forestry has raised several contradictions with the national forestry agency's powers (Singh and Kafle, 2000). Furthermore, while the initial decentralization was directly from the national to the village levels of governance, subsequent measures have introduced intermediate levels to coordinate activities and allocate funds. In addition, still lower levels of community participation were introduced in order to more effectively address sustainable forest management in an attempt to bypass the political wrangling that often characterized village units.

The experience of decentralization in Nepal highlights the difficulty that national governments face in balancing participation at a village level with allocative efficiency, horizontal equity, the role of intermediate levels of governance and the most effective level for programme delivery.

Decentralization involving customary ownership

Papua New Guinea

Virtually all forestland (97 per cent) in Papua New Guinea is owned by clan or tribal groups under customary law. The constitutionally guaranteed customary land ownership is the primary factor influencing forest use.

Successive national governments have followed a policy of using the nation's natural resources to bring development to the rural sector. First, however, the government had to acquire timber rights from the customary owners. Under the purchase process, the customary owners leased their right to exploit timber to the government for as long as 40 years. Timber harvesting and utilization rights were then granted to third parties.

Under the current Forest Management Agreement approach, the national forest authority secures the commitment of resource owners to accept recommended forest management practices, while offering investors access to the forest for a minimum of 35 years. The national government issues a permit under which it manages the forest on behalf of the customary owners for the duration of the agreement. Management is generally implemented through a developer, including harvest and construction of infrastructure. Currently there are 32 logging concession projects covering 195 Forest Management Agreement areas.

In 1995, the European Union (EU) began assisting villages (clans) in New Britain to establish small-scale logging and portable sawmilling operations that would achieve Forest Stewardship Council certification. By 1997, some 46 projects had been commenced. The process emphasizes village consultation, resolution of disputes, and planning and training. Other non-governmental organizations (NGOs) (for example, the Pacific Heritage Foundation and the World Wide Fund for Nature) have also established certification-based village projects.

The scheme provided employment and training and used traditional clan structures as a basis for decision-making, as well as generating revenues. However, there are limitations. Some are posed by the social norms: those in positions of power are expected to redistribute some of their income to clan members and to employ family, clan or regional kinsmen in preference to others, and in that order.

Fiji

The major factor determining the forest management and harvesting in Fiji is the pattern of land ownership. The bulk of the land, including productive forestland, is owned by some 6000 Fijian communities (*mataqali*). These communities do not have any corporate authority to deal in lands, and all negotiations for the use of the indigenous timber growing on their land have to be conducted through the Native Land Trust Board. Logging is allowed only by consent of both the community and the board. Timber cutting rights are then negotiated between concessionaires or licensees and the board, which authorizes the forestry department to issue logging licences, administer concession agreements and collect fees.

The revenues passed to communities are distributed under long-established formulae to a hierarchy of chiefs and, finally, to community members. The forestry department within the Ministry of Fisheries and Forests directly manages all forest reserves and indirectly manages and regulates all forest exploitation, except that of Fiji Pine Ltd.

Fiji Pine is a quasi-private 'state-owned' corporation that manages pine plantations established with national government and aid funds. The plantations are widely distributed on land owned by and leased from various communities and have been producing revenues from the sale of logs for many years. Although the communities are represented on the board of directors, dissatisfaction has developed among many of the individual communities over revenues, employment and participation in decision-making. This has led to local fire lighting, blockades against log trucks, and an unwillingness to maintain the young plantations properly. Greater devolution had been recommended (Ferguson et al, 2001); but the national government has been unable to resolve the political complexities. The present situation highlights the difficulties of collective privatization of management where land is communally owned.

DECENTRALIZATION THROUGH PRIVATIZATION

China

In 1949, China's forest had almost completely disappeared in several areas. The destruction had not only caused a drastic dearth of timber and fuel, but severe ecological deterioration. After the Peoples' Republic of China was founded, the importance of forestry as complementary to agricultural development was understood and development of forestry got a major boost. The programme to make China green was started in the first five-year plan with annual planning targets. The first national afforestation programme was set out in the 12-year National Plan for Agriculture in 1956, with emphasis on tree planting.

Responsibility for planning forest development is vested with agencies at various levels, from national to provincial (provinces/autonomous regions/municipalities), district (prefectures/cities), village (counties/towns) and household levels. The national state forest administration determines broad policy and coordinates the plans of various provinces. Provincial agencies coordinate local forestry activities. Ownership of forests falls into three categories: national (state) forests, village collective forests and private forests.

Village forest farms are spread all over the country, most in the south. They are regulated by local governments and have the responsibility for managing secondary forests, making up about 19 per cent of the nation's total forest area.

Village collective forests, making up 53 per cent of total forests, are also concentrated in southern China (Zhou, 2000). Local villages own these forests and have considerable autonomy in organizing production and generating income (Shen, 1999). Although private ownership of forest has existed for about two decades, there are no official statistics on its extent.

From the early 1980s, China initiated a series of forest policy reforms under the Household Responsibility System (Wang et al, 2004). The government decentralized power in decision-making and shifted some power to local governments and local communities, as well as to households. In 1981, the central government outlined a new strategy, urging village governments to enforce and stabilize forest tenure arrangements and encourage farmers to manage forests on a contract basis. In June 1985, national government control over timber transactions was removed, freeing the timber market.

Forestry bureaus have prepared inventories of state-owned and collective forestland where use rights could be given to private individuals, while the collectives retained ownership. If the use right is held by a co-operative, the people involved in the co-operative hold shares in the enterprise. Individuals can earn shares by contributing capital and infrastructure or through their labour in building and managing the project. Profits from the enterprise are divided among shareholders.

Since the implementation of the Household Responsibility System, private forest farming has been expanding. To date, the forest under private farming in collective forest regions accounts for 80 per cent of the total area of collective forests (Kong et al, 2002). In some provinces, private forest farming has achieved remarkable progress and played an important role in increasing farmers' incomes. Households enjoy greater freedom in decision-making and can easily respond to market signals.

Many families who received forests initially over-exploited or deforested them. But after a few years both forest area and timber stocks started to grow as farmers planted more trees. Things improved more quickly in those regions that handed over forests faster, went further towards liberalizing markets, charged lower taxes and had more consistent policies.

But this system also has critical problems – primarily economic efficiency. The fragmentation of forest ownership and management constrains the application of some technologies, such as the use of large machinery and pesticides. This may create ecological problems regarding water quality and biodiversity by fragmenting natural ecosystems (Olff and Ritchie, 2002).

Forestry development for the production of commodity products generally enjoys substantial economies of scale, and the area required for satisfactory operation is held to be relatively large. To some degree, forest management for this purpose also requires investment and a level of training and scientific competence that lies beyond the capabilities of households.

The Household Responsibility System also lacks long-term security because of the changes wrought by births, marriages and deaths. Furthermore, household partnerships tend to pursue short-term profits, to the neglect of the environmental functions of forests. At best, this option may be useful as the initial stage of private forest farming where per capita land area is limited. This stage can enable private forest owners to gain some knowledge about forestry, raise their awareness of property rights and increase their income, thus building their forest management capacity.

Vietnam

Vietnam's Renovation Policy (*Doi Moi*) of 1986 established a multi-sectoral market economy. The economy has since undergone radical reforms, one of the main elements of which has been the allocation of forestland under lease to enterprises and individual households for 20 to 50 years in order to facilitate development of the mountainous regions.

The redefined system of responsibility for forest management had three components. De-collectivization of farms to individuals started in 1986. Long-term leases were granted to farmers starting in 1988. The leases were for 30 to 50 years for lands devoted to tree crops, and 50 years for bare lands and degraded lands to be converted into forest plantations. More than 1 million hectares have been allocated to approximately 800,000 households participating in agroforestry and farm forestry programmes. The market mechanism raised prices for forest products and encouraged private investment.

Direct responsibility for day-to-day management of forests rests with the provincial and district administrations reporting to the national forestry agencies under the Ministry of Agriculture and Rural Development, which provides technical guidance and specialized services. District (state) forest enterprises constitute the main basis of the organizational system in forestry. There are 413 forest enterprises controlling about 4.7 million hectares of forests, of which 2.8 million hectares are natural forests: 240 district enterprises are involved in managing production forests.

However, scattered natural forests are also allocated to households, communes/co-operatives and villagers. Contracts with households are usually 50 years and cover ecological restoration, afforestation and protection (Phuong, 2000). Some 466,000 households have received protection contracts covering an area of 1.6 million hectares. Production contracts covering 50 per cent of the production forests have gone to 473,500 households and the rest to 7442 co-operatives and enterprises. Allottees of forestland are entitled to receive credit for investment, or cost sharing on an equal basis.

The forestland allocation system has encouraged agroforest farms and tree plantations and raised household income. However, the rights and benefits of households are not well defined, and this discourages some farmers from investing in forest production. Furthermore, the choice of species and business decisions are still highly centralized and may not be in the best economic or social interests of farmers (Woods, 2003). Finally, many of those who rely on the forests for all or part of their livelihood, especially indigenous groups, are landless and cannot participate.

Australia

One way in which Australia has sought to encourage private-sector investment in plantations is to allow producers to charge the costs of establishment against taxable income at the time when those costs are incurred, rather than waiting until the timber revenues are realized (see Gregersen et al, 2004). Another is to allow the raising of pooled funds from small investors through prospectuses and their investment in plantations.

The national and provincial (state) governments also developed the National Competition Policy relating to all commercial activities owned by provincial governments. This agreement requires all government commercial activities to be competitively neutral (no cross-subsidies). This change coincided with a general shift towards commercialization, corporatization and privatization in the provincial forestry agencies.

To date, two provinces (Victoria and Tasmania) have privatized their softwood plantations, a third (New South Wales) is likely to do so soon, another (South Australia) has formed a state-owned corporation, and the remainder have commercialized their operations. These changes have increased the return on investment and reduced direct forestry employment in a country with high labour costs. Investment in hardwood plantations has grown very rapidly during the last 15 years (Ferguson et al, 2003), much of it through managed investment schemes.

All plantations, public or private, are subject to legislated codes of forest practice in order to ensure that operations are environmentally sensitive. Many of the larger companies are seeking certification, principally because of shareholder and stakeholder pressures, but also to improve access to export markets. Certification also involves checks and restraints on the sustainable yield, a matter that is generally not clearly dealt with in codes of practice. This is important in the case of takeovers because of the temptation for new owners to recoup some of the investment by over-cutting. The plantation sector has demonstrated the capacity of privatization to achieve greater allocative efficiency while maintaining and improving environmental standards, and to create a favourable environment for appropriate investment in growing and processing.

Some issues remain, however, such as the concerns of some communities about radical changes in land use. These will continue to be the focus of community, district and, sometimes, provincial debates.

Reversal of devolution in a federal system

Australia

Under Australia's constitution, administration of all matters of land management, including forestry, were delegated to the provinces. This arrangement worked well until the 1950s: almost all public concerns about forest management were confined territorially within provinces, and the provincial forestry agencies had considerable control over the publicly available knowledge base. Given the history of uncontrolled fire lighting and uncontrolled and sometimes illegal logging, this command-and-control management was probably appropriate for the times.

Since the 1950s, the public has become well informed about environmental matters, and the provincial land management agencies have struggled to deal with the resulting shift to greater public participation. These changes have increased the policy role of the national government in forestry matters. Any substantial dispute in the field of forest conservation and the environment quickly became a national matter through the combined influences of the media, especially television, and the national stakeholder groups (principally the forest industry, unions, landholder and NGOs).

Increasingly, the national government has used its powers to influence outcomes at a provincial or district level, often overriding or confounding the actions at the lower level, particularly through its control of exports and of World and Australian Heritage matters. Where the national and provincial government concerned were of different political persuasions, both sides frequently provoked deliberate disputation, especially over the granting of export licences by the national government. In 1992 a massive public protest by stakeholder groups culminated in a temporary blockade of Canberra by logging trucks. Subsequently, a joint national–provincial agreement (the Regional Forest Agreement process) was developed.

Today, most policy outcomes appear reasonably stable. Nevertheless, conservation groups, while welcoming the increase in conservation reserves that the joint agreement has brought, have renewed the debate over areas not included. Two provinces have already reacted by unilaterally increasing the conservation reserves, thereby raising questions about national government versus provincial authority in a federal system.

The situation in Australia also invites comparison with that in the US, with parallels in old-growth forests, harvesting of native forests, wildlife conservation and hydrology. Yet the US has strong centralized control over national forests and national parks, which are major foci for environmental concerns (compare Chapter 14, which stresses state diversity). Decentralization (or the partial reversal of it in Australia) is, thus, neither a necessary nor a sufficient condition for addressing these debates.

Charting future paths and avoiding pitfalls

The current manner of democratic control can best be described as a polyarchy (Dahl and Lindblom, 1953): a system in which non-leaders can exercise a high degree of control over leaders through voting. A polyarchy typically involves several overlapping subsystems, including those dealing with governance, bureaucracy, markets and information.

Governance

No democracy can function effectively without the rule of law. The legal bases of delegation and devolution need to be fully developed and understood. Although decentralization can delegate and devolve authority, it cannot completely delegate or devolve responsibility. Higher levels of governance and bureaucracy always retain some responsibility for lower levels, if only as the last resort when things go wrong at lower levels.

The art of decentralized governance is therefore to strike an effective partnership between the respective levels of governance involved and the bureaucracy used for different functions and situations. A variety of modes of decision-making can be used: majority-rule decision-making, consensus through veto, consultative processes by third parties, consultative processes by the bureaucracy and direct bureaucratic decision.

Much of formal democratic governance is based on voting and majority rule. In exercising that majority rule, majority leaders cannot completely ignore the minority or civil rule may become unstable over deeply felt divisions, such as those often associated with ethnic and environmental concerns. Hence, leaders typically establish trade-offs with minority groups to obtain their support, albeit qualified. The process inherently involves and embraces conflict. Conflict leads to political action in a democracy, the promotion of information about the issue, and a better informed public and elected representatives to vote on it. Conflict through public debate is therefore not necessarily counterproductive and may be very important in promoting democratic governance.

Seeking consensus among local stakeholders is, nevertheless, often espoused as a desirable characteristic of decentralization and, by inference, of greater democratic control. Any one group of stakeholders, however, can exercise a veto in seeking consensus. Coglianese (1999) points out that this places the central government in a different situation – that of being a facilitator of decisions or, at most, of being 'just another player in the game'. Those who are best represented in the political process tend to become dominant in policy-making, possibly to the detriment of policy-making in the public interest. Although a consensus through veto may be acceptable if there are few stakeholder groups that encompass virtually all interests, it is not appropriate for those in which, in addition to strong stakeholder groups, there are widely dispersed and/or poorly represented interests. Furthermore, some evidence (Purnomo et al, 2004) is emerging that local stakeholder perceptions of sustainability are at least in one case different (statistically and otherwise) from those of NGOs, industry and public forestry agency employees.

To the extent that reducing conflict is a desirable goal (Fung and Wright, 2003), consensus through veto may be useful but has disadvantages. Inequalities between groups may lead to domination or even capture of the process. Groups may participate only when it suits them. In espousing consensus through veto, elected and appointed officials may simply be protecting their own interests for either electoral support or a quieter administrative life. The approach may reduce information for the public at large because the interest groups are reluctant to promote the negatives to them that have had to be traded. This approach is even said to lead to the 'lowest common denominator' solution to the possible detriment of the public interest at large.

As Coglianese (1999) points out, decentralized policy-making *can* be more efficient when based on a participatory process in which the administrative agency simply consults with local people about the issues before drafting policy and then tests reaction to it. The many forms such a process can take have been summarized for forest management by Mayers and Bass (1999).

Where substantive delegation or devolution from central (or provincial) government to local government is likely to be involved, the territorial stakeholders need to be well represented. Where the decentralization changes are directed at particular communities and problems, rather than the nation, the representation needs to incorporate links to existing local government structures, and those structures need to be included in capacity development. Nothing undermines devolution arrangements faster than the perception that a new body is usurping existing power.

Corruption represents a major impediment to sustainable forest management, with or without decentralization. Petty corruption in forest management can be curtailed by instituting appropriate measures if, and only if, major corruption at the level of national or sub-national governance is addressed and curtailed (Callister, 1999). This is a matter of political will that requires a moral strength of leadership and the support of the population at large.

Bureaucracy

If governance is decentralized, then the previous central bureaucracy will need to be deconstructed in some way. While the geographic structure of the bureaucracy tends to mirror that of governance – the national, provincial, district, village, household hierarchy – other structures and arrangements are possible and sometimes desirable.

Particular issues that cross local, regional or national boundaries may require the formation of collective institutions. This may lead to delegation of some powers upwards. Examples include collectives of adjacent local governments for managing a forest resource common to their boundaries, and collectives of adjacent provincial agencies and national agencies to manage a major water resource common to all.

Examples of private and quasi-private corporations are numerous in the electricity and water field. Where publicly owned native forest resources are involved, quasi-private organizations in the form of state-owned corporations are more typical than purely private corporations. Quasi-private organizations have also been tried in the management of national parks. These operate commercially through contractual agreements with the administering government department to supply services at annually negotiated quantities and prices.

Decentralization inevitably threatens the vested interests of the established bureaucracy because it entails geographic moves for some and reductions in staff for others. This is a serious problem in all countries with strong central bureaucracies (see, for example, Khan, 1998). The extent of consultation and participation of the bureaucracy may therefore have to be limited and external sources used in designing decentralization. External monitoring of subsequent outcomes also becomes essential.

Yet bureaucracy also needs assistance in coping with the changes posed by delegation or devolution. Insufficient resources, administrative systems and training invite poor outcomes and corruption. Without these, accountability becomes a fiction.

Finally, the linkages between elements of bureaucracy also need to be considered in any decentralization. Sustainable forest management and poverty reduction programmes involve the integration of certain functions across departments or agencies. These need to be taken into account in designing delegation or devolution of functions.

Markets

Market-based approaches to forest use encourage entrepreneurial behaviour, spontaneously correcting errors and enhancing the mobility of resources and the value of tradeable assets. Privatization of plantation rights has improved efficiency and created a new asset class recognized by commercial investors.

Markets can potentially play a much wider role if a more detailed approach is taken in defining rights. Ownership of rights can be unbundled: public ownership of land can be retained while the timber resource is privatized. This facilitates devolution of harvesting and silviculture to local bodies operating under contracts that require them to conform with nationally or regionally set codes of practice, sustainable yields and management plans. In this way, minimum standards of environmental care can be maintained and even improved.

The question is how much of the regulation of those standards can be devolved to local communities. Where corruption in the bureaucracy is an issue but common property rights exist, inherent incentives exist for local communities to regulate effectively; local and indigenous communities have come to regard the protection of forest values as important and a matter in which they wish to participate in determining the balance of joint production.

Much depends upon education, knowledge and levels of interest in the local community. Rights to harvest timber, for example, can be subjected to the competitive discipline of allocation and price determination through appropriately designed auctions and tenders. Other private property rights relating to timber harvesting need to be clearly defined; provide for transferability; include sanctions to ensure adherence to codes of practice; include ratchet provisions in the case of long-term licences that progressively diminish the annual volume harvested, enabling increasing sales on spot markets; provide for periodic competitive renewal and redetermination of price; and include bond provisions as a protection against environmental damage and default.

Information

Most aid projects have paid insufficient attention to information provided to stakeholders at large. Radio is the cheapest and most widely available way of disseminating information in developing countries. Local radio companies in rural areas need information packages that they can revamp for their listeners.

But the media are not the only means of sharing information. Public meetings, political parties, other collective organizations and networks, and even opinion surveys can also be used. Formal grievance procedures represent yet another source of information collection and promulgation and can contribute to the development of informed debate and decisions.

A well-functioning local democracy should have an array of information sources. Although a multiplicity of conflicting sources can be confusing, the persistence of the electoral process and debate normally ensures appropriate change. This illustrates the role of the 'science of muddling through' that Lindblom (1959) identified as one of the likely features of democratic policy-making under conditions of low understanding and major change.

Formal education programmes, whether short or long, involve a longer-term and deeper development of knowledge and skills and are therefore vital in capacity-building. But the accountability of bureaucracy to elected officials and the accountability of elected officials to the citizenry remain essential (Blair, 2000).

Lessons learned

The lessons learned from the Asia Pacific experience can best be summarized under the headings of preparation, process, devolution and privatization.

Preparation

The objectives of decentralization are good governance marked by improved efficiency and equity, transparency, accountability and people's participation; balanced and sustainable development; and empowerment of the people. Achieving an appropriate balance between empowerment of people at different geographic levels or in different local communities is the issue.

Decentralization is not a panacea, nor is it always efficient or equitable. It can improve democratic governance and, in doing so, may assist poverty alleviation and sustainable forest management; but it is not a sufficient measure. Decentralization is a long-term process to be accomplished in phases.

Decentralization does not mean doing away with controls. It tends to be more successful under a strong central government than under a weak one. It is often constrained by a tendency for centralization of what was decentralized. It may be more difficult in highly stratified societies with strong ethnic, caste, tribal, income and gender prejudices.

Careful design and development of the legal basis of decentralization is a desirable precursor to change in order to ensure that the rights and responsibilities are clearly defined and based on adequate consultation, sanctions, grievance procedures and systems to implement and monitor them.

Decentralization measures, especially those involving devolution, need to address systemic corruption in existing and potential new structures before initiating change in order to ensure that corruption is not simply shifted to other levels of government – at the cost of continued poverty and unsustainable forest management.

Process

In a formal sense, democratic governance generally rests on the majority-rule principle. However, an array of other forms of decision-making exist, such as consensus by veto, consultative processes through third parties or the bureaucracy, and direct bureaucratic action in the case of natural disasters. The forms chosen also need to recognize the historical and cultural setting.

The bureaucracy is itself a potential power elite. Devolution without the active support of the bureaucracy is difficult to achieve successfully. Devolution to lower levels introduces principal–agent relationships, requiring incentives as well as sanctions to ensure that the agent acts in accord with the principal's goals.

Where regulation is separated administratively from delegated or devolved forest management, regulatory functions should generally be conducted at least one level of the bureaucratic hierarchy above the latter in order to provide sufficient independence from local politics and power elites.

Collective bureaucratic units may be needed to deal with transboundary issues or provincial and national issues in a federation and deserve special attention in terms of their legal basis to achieve an effective partnership.

Information is vital to the functioning of a modern democracy and more attention needs to be given to promoting information about decentralization goals and strategies through the media and other means of dissemination in order to assist conflict resolution. The increasing prominence of national media may lead to a partial reversal of decentralization, especially in federal systems.

The positive role of conflict resolution through public debate needs to be recognized and fostered, especially where majority-rule decision-making is involved. Consensus by veto approaches should be confined to those situations in which there are only a few stakeholders of comparable strength. In some situations, consultative processes by third parties or the bureaucracy, and even direct action, may be more appropriate than either of the former modes.

Devolution

In addition to providing a proper legal basis and resources, devolution to lower levels has to involve a meaningful transfer of authority in order to be acceptable and work effectively. It also has to maintain horizontal equity by ensuring a fair distribution of fiscal and other resources across the units. Too much decentralization (autonomy) may lead to neglect of the overall national picture and failure to implement nationally planned priorities. Successful efforts can enhance participation, increase the regional share of income from forests, result in better delivery of services and improve the sustainability of forests.

Devolution to village levels is more likely to be effective when adapted to traditional systems of governance and when aimed principally at poverty alleviation and fuelwood supply. When aimed at commercial wood production, the scale of operation may pose an impediment and require the formation of collectives of village units.

Devolution to customary ownership units is necessary where land or resource ownership has traditionally been customary. The formation of collectives may then have to be encouraged in order to provide commercial scale of wood production, leading to sensitive issues of representation in their governance. Small-scale logging and sawmilling can be operated through customary units but require substantial aid, support and a cost-effective framework for certification.

Privatization

Privatization represents an efficient and self-regulating form of decentralization in relation to the supply of commercial forest goods and services and tends to be conducive to private investment in the sector. But forest production also often involves the supply of non-market goods and services, including those of environmental protection. The legal basis of property rights for private and quasi-private forest management units therefore needs to be well defined, and to include opportunities for participation by all stakeholders in developing the rules, as well as providing for sanctions and grievance procedures.

Note

Ian Ferguson acknowledges a vested interest as a director of a softwood plantation company in Australia and chair of another. This chapter was commissioned by the International Tropical Timber Organization (ITTO); however, the views expressed here are those of the authors and do not necessarily reflect the views of ITTO or its members.

References

Agrawal, A. and Ostrom, E. (2003) 'Collective action, property rights and decentralization in resource use in India and Nepal'. *People and Society*, vol 29, no 4, pp485–514

Blair, H. (2000) 'Participation and Accountability at the Periphery: Democratic

Governance in Six Countries', *World Development* 28(1): 21–39
Brack, D., Marijnissen, C. and Ozinga, S. (2002) *Controlling Imports of Illegal Timber: Options for Europe*. Royal Institute of International Affairs and FERN, London
Brown, D. W. (1999) *Addicted to Rent: Corporate and Spatial Distribution of Forest Resources in Indonesia: Implication for Forest Sustainability and Government Policy*. Indonesia–UK Tropical Forest Management Programme, Provincial Forest Management Programme, Report no PFM/EC/99/06, Jakarta
Brunner, J., Seymour, F., Badenoch, N. and Ratner, B. (2000) *Forest Problems and Law Enforcement in Southeast Asia: The Role of Local Communities*. World Resources Institute Resources policy brief, Washington, DC
Callister, D. J. (1999) *Corrupt and Illegal Activities in the Forestry Sector: Current Understandings, and Implications for World Bank Forest Policy*. Draft report for discussion, World Bank Group, Forest Policy Implementation Review and Strategy Development: Analytical Studies, Washington, DC
Chandrasekharan, C. (2003) *Policy and Institutions: Development in the Philippine Forestry Sector*. Consultancy report UNDP.SPPD/PHI/01/010, UNDP, New York
Chen, K. and Brown, C. (2001) 'Addressing shortcoming in the household responsibility system: Empirical analysis of the two-farmland system in Shandong Province'. *China Economic Review*, vol 12, no 4, pp280–292
Coglianese, C. (1999) 'Rethinking consensus: Is agreement a sound basis for regulatory decisions?' Paper presented at conference on Environmental Contracts and Regulation: Comparative Approaches in Europe and the United States, University of Pennsylvania Law School, 24–25 September
Dahl, R. A. and Lindblom, C. E. (1953) *Politics, Economics and Welfare*. Harper and Brothers, New York
Dangol, S. (2005) 'Participation and decisionmaking in Nepal', in Colfer, C. (ed) *The Equitable Forest: Diversity, Community and Resource Management*. Resources for the Future, Washington, DC, pp54–71
Dauvergne, P. (1997) *Shadows in the Forest: Japan and the Politics of Timber in Southeast Asia*. MIT Press, Cambridge, MA
Esman M. J. and Uphoff, N. T. (1982) *Local Organizations and Rural Development*. Cornell University, Rural Development Committee, Ithaca, NY
FAO (United Nations Food and Agriculture Organization) (1982) *Village Forestry Development in the Republic of Korea: A Case Study*. FAO, GCP/Int/347(SWE), Rome
Ferguson, I. S. (1996) *Sustainable Forest Management*. Oxford University Press, Melbourne
Ferguson, I., Fox, J., Baker, T., Stackpole, D. and Wild, I. (2002) *Plantations of Australia: Wood Availability 2001–2044*. Consultant's report for the National Forest Inventory, Bureau of Rural Science, Canberra
Ferguson, I., Leslie, A., Pens, H., Reid, R. and Shepherd, P. (2001) *Review of Fiji Pine Industry: Final Report: 23 November 2001*. Department of Forestry, University of Melbourne, Melbourne
Fung, A. and Wright, E. O. (2000) 'Deepening democracy: Innovations in empowered participatory governance'. *Politics and Society*, vol 29, no 1, pp5–42
Furtado, X. (2001) *Decentralization and Capacity Development: Understanding the Links and the Implications for Programming Capacity Development*. CIDA Policy Branch Occasional Paper Series no 4, CIDA, Quebec, Canada
Gershberg, A. I. (1998) 'Decentralization, recentralisation and performance accountability: building an operationally useful framework for analysis'. *Development Policy Review*, vol 16, pp405–431

Gregersen, H., Contreras-Hermosilla, A., White, A. and Phillips, L. (2004) *Forest Governance in Federal Systems: An Overview of Experiences and Implications for Decentralization*. Draft Report circulated at the Interlaken Workshop on Decentralization in Forestry, 27–30 April 2004, Interlaken, Switzerland

Grindle, M. S. (2000) *Designing Reforms: Problems, Solutions, and Politics*. Harvard University, Kennedy School of Government, Faculty Research Working Papers Series, November

Hobley, M. (1996) *Institutional Change within the Forestry Sector: Centralised Decentralisation*. Overseas Development Institute, Working paper 92, London

Kasa, S. (1999) 'Political power and the Indonesian forest concession system', in Palo, M. and Uusivuori, J. (eds) *World Forests, Society and the Environment*. Kluwer Academic Publishers, Dordecht

Khan, N. A. (1998) *The Political Economy of Forest Resource Use*. Ashgate Publishing, Aldershot, UK

Klugman, J. (1994) 'Decentralization: A survey of the literature from a human development perspective', UNDP occasional paper 13, UNPD, Human Development Report Office, New York

Kong, M., Liu, C., Zhang, X. J. and Li, Y. M. (2002) *Research on Policy Problems Affecting the Development of Joint Stock Partnership Arrangements in China*. Unpublished report, China National Forestry Economics and Development Research Centre, China

Lee, D. K. and Kwon, L. Y. (2002) 'Roles of *Saemaul Undong* in reforestation and NGO activities for sustainable forest management in Korea'. Paper presented at the International Workshop on Forest Science and Forest Policy in the Asia Pacific Region, Building Bridges to a Sustainable Future, IUFRO/USFS/MSSRF, 16–19 July, Chennai, India

Lindblom, C. E. (1959) 'The science of muddling through'. *Administrative Review*, vol 19, pp79–89

Litvack, J., Ahmad, J. and Bird, R. (1998) *Decentralization in Developing Countries*. World Bank Sector Studies Series, Washington, DC

Lu D. (2004) *Private Forest Farming in China*. Masters thesis, University of Melbourne, Melbourne

Mayers, J. and Bass, S. (1999) 'Policy that works for forests and people'. *Policy That Works for Forests and People Series*, No 7, International Institute for Environment and Development, London

Michels, R. (1911, 1968) *Political Parties*. The Free Press, New York

Olff, H. and Ritchie, M. E. (2002) 'Fragmented nature: Consequences for biodiversity'. *Landscape and Urban Planning*, vol 58, pp83–92

Phuong, P. X. (2000) 'People's participation in forest management in Vietnam', in Enters, T., Durst, P. B. and Victor, M. (eds) *Decentralization and Devolution in Asia and the Pacific*. RECOFTC report no 18 and RAP publication 2000/1, Bangkok

Purnomo, H., Mendoza, G. A. and Prabhu, R. (2004) 'Analysis of local perspectives on sustainable forest management: An Indonesian case study'. Submitted to *Journal of Environmental Management*, vol 74, pp111–126

Rondinelli, D. (2002) *What Is Decentralization?* World Bank Institute, Intergovernmental Fiscal Relations and Local Financial Management Programme, Fiscal Decentralization, Briefing Notes, www.worldbank.org/wbi/publicfinance/documents/Topic01_BN1.pdf

Ribot, J. C. (2002) *Democratic Decentralization of Natural Resources: Institutionalizing Popular Participation*. World Resources Institute, Washington, DC

Rufi'ie (2002) *Competitive Bidding for Forest Concessions in Indonesia*. PhD thesis, University of Melbourne, Melbourne

Rukmantara (2004) 'Repairing the road to SFM'. *Tropical Forest Update*, vol 13, no 4, pp 6–7

Sève, J. (1999) *Some Views on Decentralization in a Unitary State and Implications for Community-Based Forest Management: Lessons from France*. Natural Resources Management Program Discussion Paper, Jakarta

Shen, W. X. (1999) 'Property rights and reform of forest economic system'. *Forest Economic Issues*, vol 5, pp15–19

Singh, H. B. and Kafle, G. R. (2000) 'Community forestry implementation: Emerging institutional linkages', in Enters, T., Durst, P. and Victor, M. (eds) *Decentralization and Devolution of Forest Management in Asia and the Pacific*. FAO/RAP, Bangkok

Sustainable Ecosystems International Corp (SUSTEC) (2001) *Review of Forestry Policies vis-à-vis the Watershed and Ecosystem Management Framework and the CBFM Strategy and Programme*. Consultancy Report, Manila

Suwondo, K. (2002) 'Decentralization in Indonesia'. Paper prepared for the International Non-government Organization Forum on Indonesian Development (INFID) Annual Advocacy 2002, www.infid.be/INFID%20Background%202002%20Decentralization.pdf

Turner, M. and Podger, O. (2003) *Decentralization in Indonesia: Redesigning the State*. ANU Press, Canberra

Wang, S., van Kooten, G. C. and Wilson, B. (2004) 'Mosaic of reform: Forest policy in post-1978 China'. *Forest Policy and Economics*, vol 6, pp71–83

Woods, P. (2003) *Spontaneous Agroforestry: Regreening the Barren Hills in Vietnam*. PhD thesis, University of Melbourne, Melbourne

Yoo, B. Young Il (1997) *Indepth Study on the Republic of Korea: Status, Trends and Prospects to 2010*. Asia Pacific Forestry Sector Outlook Study, working paper APFSOS/WP/06, FAO, Rome and Bangkok

Zhang, A. M. and Jia, Q. (2001) 'Discussion on reform of state forest resource management system'. *Forest Economics*, vol 5, pp18–19

Zhang, J. R. and Chen, X. H. (2001) 'Review and prospects of China's forest reform. Paper presented at the International Symposium on the Lessons from Chinese Forest Policy Experience, 20–23 June, Dujiangyan, Sichuan, China

Zhang K. (2000) 'Issues relating to the reform of forest management in China', in Enters, T., Durst, P. B. and Victor, M. (eds) *Decentralization and Devolution in Asia and the Pacific*. RECOFTC Report No 18 and RAP Publication 2000/1, Bangkok

Zhang, Y. (2001) 'Institutions in forest management: Special reference to China', in Palo, M., Uusivuori, J. and Mery, G. (eds) *World Forests, Markets and Policies*. Kluwer Academic Publishers, Dordecht

Zhou, X. N. (2000) 'Study on Private Forest'. *Forestry Development*, vol 5, pp6–14

Chapter 4

Choosing Representation: Institutions and Powers for Decentralized Natural Resources Management

Jesse C. Ribot

INTRODUCTION

Decentralization reforms are taking place in most developing nations, transforming the local institutional infrastructure upon which local forest management is based. Although intended to produce institutional conditions for more equitable and efficient resources management, most decentralization reforms fail to establish the basic conditions that theory predicts would lead to positive outcomes. This chapter argues that the central condition for effective decentralization is *representation*. Democratic representation consists of downwardly *accountable* and *responsive* local authorities. Accountability requires that the local population can sanction the local authorities via various accountability mechanisms. Responsiveness requires powers that enable these local authorities to respond to local demands. Because of resistance by governments and poor choices by non-governmental organizations (NGOs) and donors, local institutions that are not accountable to the local populations are often being chosen for 'decentralized' forest management. In addition, they are often given inappropriate and insufficient powers. This chapter asks which local institutions best serve the function of representation in decentralization reforms. It also queries the choice of powers being devolved to local institutions in natural resource decentralizations. These two elements of decentralization cannot be separated: accountability without powers is empty (there is no responsiveness), and power without accountability is dangerous (there is no check on the exercise of power). The chapter concludes with questions that can help policy-makers, donors and researchers evaluate whether decentralization efforts are being legislated and implemented in a manner that theory indicates will result in positive outcomes.

REPRESENTATION AND THE BENEFITS OF DECENTRALIZATION

Development agents, natural resource managers and many environmentalists believe that decentralization can increase both efficiency and equity in natural resources management (Ribot, 2002a). Decentralization may also be a way of leveraging the development of local democracy. What local institutional base will result in effective decentralization of forest management and strengthen local democracy?

Theory implies that the efficiency and equity benefits of decentralization come from the presence of 'democratic' processes that encourage local authorities to serve the needs and desires of their constituents (see Ribot, 2004). Following Moore (1998), this chapter defines 'democratic' substantively as the accountability of leaders to the people. The developmentalist logic behind decentralization – that is, decentralization for the purposes of local and national development – is that democratic local institutions can better discern and respond to local needs because they have better access to information and are more easily held accountable to local populations (see Ribot, 2002b). For local people to want to engage these authorities (to hold them accountable), the powers they hold and the services they can deliver must also be relevant to local people. Furthermore, these authorities must have some freedom of decision-making in order to be responsive in a way that reflects local mandates. The power to implement mandates handed to them from above is not sufficient. They must also wield discretionary powers. Without discretion local authorities may be able to implement imposed agendas; but to implement local mandates, the authorities need flexibility. In brief, theorists believe that effective decentralization requires representative authorities with discretionary powers over resources that are meaningful to local people (Ribot, 2002b, 2004).

In the decentralization reforms taking place today, a limited array of forest management powers are being transferred to a wide variety of institutions – including appointed and elected local government authorities, local forest service agents, NGOs, user groups and traditional authorities. The experiences from community-based forms of natural resources management indicate that democratic local institutions can be the basis of effective local environmental decision-making, that communities have or can develop the skills and desire to make and effectively execute natural resources management decisions, and that community-level management can have ecologically and socially positive effects. Experience from decentralization indicates that local communities can manage decentralized natural resources well (Larson and Ribot, 2005). Outcomes depend on a variety of regulations and incentives that shape use patterns.

In short, the outcomes of decentralization seem to be a function of representation – which is composed of powers and downward accountability – and management incentives. It is clear that local people can manage their natural resources under certain circumstances. The following sections focus on the implications for representation of institutions being chosen in natural resource decentralizations and sketch a theoretical model of representation against which to examine decentralized natural resources management schemes.

Representation: A model

Representation is key to the decentralization formula. Following Manin et al (1999, p2), it is useful to break representation into *responsiveness* and *accountability*. They model the policy process as a chain in which preferences expressed through various signals become mandates and are translated into policies and then outcomes. Responsiveness is the relation between signals and policies – the ability of decision-makers to deliver. Accountability is the relation between outcomes and sanctions – the ability of people to make demands and threats. A government is 'responsive' if it adopts policies that are signaled as preferred by citizens. Governments are 'accountable' if citizens can sanction them appropriately. A government is 'representative' because it is responsive and/or accountable.

The accountability relation is established through an ensemble of sanctions or 'accountability mechanisms'. Accountability is a counter-power – that is, any power that balances or puts a check on the power of other power holders (Agrawal and Ribot, 1999). Accountability is constituted by the set of mechanisms and sanctions that can be used to ensure that policy outcomes are as consistent with local needs, aspirations and the best public interest as policy-makers can make them. Responsiveness is a function of the many factors that enable local authorities to translate local needs and aspirations into policy. These include the appropriate mix of powers to act on behalf of the people and the abilities to analyse and to translate signals into policy. Responsiveness is also a matter of will. It may be motivated by the fear of sanctions or it may be a function of ideology – that is, public spiritedness or a belief in public service. What makes democratic systems unique is that they do not rely upon ideology alone to ensure that the full policy cycle is representative. Democratic systems rely upon a mix of ideology and sanction, with the sanction as the guarantee (Manin et al, 1999, p2).

This model suggests several important dimensions for comparing the appropriateness of local institutions as the recipients of decentralized powers. The accountability side of the equation indicates accountability mechanisms, such as elections, magic, sabotage and protest. The responsiveness side of the equation indicates that powers are important: external powers, such as finances and the ability to mobilize resources and labour; and internal powers, such as the capabilities and knowledge needed to exercise external powers in translating signals into effective policies. In the external domain are the classic executive, legislative and judicial powers, including extended powers such as the ability to mobilize central government to deliver services, technical assistance, equipment and finances, which can be exercised through legal rights or political pressures, social relations and so forth. In short, the primary elements of representation can be boiled down to accountability and power.

Accountable actors with powers – the 'actors, powers and accountability model' – provides a useful, but static model (Agrawal and Ribot, 1999; Ribot, 2004). Effectiveness has temporal and spatial dimensions as well. Sustainability of representation over time and its replicability or ability to be scaled up also matter. Sustainability and replicability need to be further theorized and developed as concepts: how do we choose representative institutions that can also be sustained and scaled up? Certainly there are different bases of sustainability. Customary authorities have endured – with greatly varying degrees of legitimacy – over centuries. NGOs, private voluntary organizations (PVOs) and community groups come and go as a function of local need, finance and external intervention. Local governments are only as sustainable as the central state that legislates them into existence. Traditional authorities, however (even those that are stable and representative), are difficult to replicate elsewhere. Elected local government, in contrast, can be legislated across a national territory.

Furthermore, while ideology may drive some actors to be representative, the absence of accountability mechanisms allows drift towards self-interest and power consolidation. So, although representation can occur without accountability, it is highly questionable whether it can be sustained without systematic forms of accountability. Lastly, sustainable representation may rely as much upon the powers and accountability of individual institutions as it relies upon the particular mix and hierarchy of institutions embedded in their particular place and history. This observation raises the question of how actors external to the local arena – NGOs, governments and donors – should view institutional hierarchies and the accountability relations among local institutions.

In sum, the important institutional dimensions for effective representation are powers and accountability mechanisms, and the degree to which these are sustainable and are able to be scaled up. At a minimum, it appears that if local decision-makers are to act independently on matters important to local people, their mix of powers must be sufficient to the mandates, meaningful to local people and discretionary. The accountability mechanisms must be systematic and effective. In addition, institutions being empowered should be sustainable over time and replicable across territorial space. Which local institutions are most likely to have these characteristics is an empirical question.

Making choices

Legislating and implementing decentralization are the first steps. But even where secure decentralization has been implemented, support from central government and others are needed to ensure that natural resources are not over-exploited, that equity is not compromised and that legislation and implementation do not work against each other. Some of these efforts include minimum environmental standards and alleviating poverty, as well as accompanying measures for civic education and conflict mediation. Central government must play a key role in advancing reforms needed to achieve effective decentralization. In practice, an end-point of decentralization reform is never reached, since reform entails an ongoing political struggle between local and central interests.

Institutional choice

Central ministries are allocating powers to a variety of local institutions in the name of decentralization. In order to evaluate whether the choice of local institutions will lead to effective decentralization, the key question is whether the selected institutions are accountable to the populations for whom they are making decisions. Often they are not, since effective decentralization is not the only purpose of those choosing local institutions. Central authorities depend upon local institutions for implementing central agendas, legitimizing state projects, incorporating breakaway groups and regions, garnering popular support, obtaining an electoral base, cultivating patronage networks and so forth. International donors and NGOs depend upon local institutions for implementing their specific environmental, health, educational and infrastructure agendas – whether or not local people are interested (Ferguson, 1996; Schroeder, 1999; Baviskar, 2002, 2004). Local and national elite also have interests in capturing and using local institutions and the powers they are receiving. Faced with these powerful competing interests, locally accountable and representative institutions are often sidelined. Because of these countervailing forces, choosing and building on representative and accountable local institutions is a critical aspect of decentralization.

Electoral accountability can be strong or weak, depending upon the electoral process. India, Mali, Uganda and Mexico have chosen to strengthen local accountability by admitting independent candidates in local elections (Agrawal, 2001; Baviskar, 2002; Bazaara, 2002a; Dupar and Badenoch, 2002; Farrera, 2002; Kassibo, 2002a). In Burkina Faso, Cameroon, Senegal, South Africa, Zimbabwe, Brazil, Bolivia, Nicaragua, China, Laos, Indonesia and Vietnam, however, local elections take place by party list. The elected authorities are often more accountable to their parties than to the local population (Ribot, 1999a; Dupar and Badenoch, 2002; Resosudarmo, 2002). In Bolivia, some councils have acted on behalf of local populations in keeping timber concessions out of their forests, while others allowed them to operate locally, despite popular opposition (Pacheco, 2004). In places where only the party in power has the means to organize candidate lists across the country and there is no real competition among parties, these systems leave little chance for local populations to choose their own representatives (Ribot, 1999a). But even where there are elected local governments, central governments and donors often avoid them in favour of other kinds of local organizations. In many instances governments, donors and NGOs avoid local elected bodies as being too 'political' or as being inefficient or lacking in capacity (Ferguson, 1996; Baviskar, 2002). It is the very political nature of local elected bodies that makes them accountable to local needs and aspirations (Romeo, 1996, p4; Evans, 1997). In addition, elections may not strengthen environmental accountability where natural resources are not a central issue (Dupar and Badenoch, 2002; Larson, 2004). When locally accountable and surrounded by a plurality of voices, elected institutions can serve as an integrative mechanism for local decision-making (Ribot, 2001a).

Other groups in the local arena are often empowered in decentralization reforms. These include central government administrators or line ministries, membership organizations, NGOs, single-purpose committees and user groups and customary authorities (see Larson and Ribot, 2005). How democratically accountable are these institutions to local populations? The current wisdom in democratic decentralization is that for management of public resources such as forests, pasture lands and fisheries, accountability should run from these groups through elected local bodies to the people (Blair, 2000). However, these non-elected organizations are often empowered as if they are themselves representative or democratic, which they often are not. Even though local governments may not always be democratic, these alternative institutions have even less systematic accountability to the public at large. Local accountability of elected local governments may increase if both groups – empowered non-elected groups and elected government – are monitored and offered assistance by deconcentrated central government offices, are surrounded by interest groups and NGOs and are faced with active customary authorities in the local arena. But empowering such institutions in place of elected authorities can be anti-democratic.

Deconcentration to *local branches of central ministries* is not very different from decentralization to upwardly accountable, party-selected local representatives. Deconcentrated institutions lack some of the local accountability that is believed to make decentralization work. Nevertheless, there is evidence that deconcentration can serve local interests well. In Brazil, for example, a system of performance awards led civil servants to better serve local needs (Tendler, 1997). Local line ministry offices can support local democratic authorities, but should not substitute for them because such substitution can de-legitimize and undermine the stronger form of democratic decentralization.

Grassroots groups and *NGOs* may also not be accountable to, or representative of, local people in a systematic manner. Rather, they are constituted to represent the interests of their members. In addition, the internal democracy of grassroots groups and NGOs is not assured. Transferring powers to these organizations cannot be considered more democratic or representative than privatization – which is not a form of decentralization.

Spokespersons for various local movements or organizations are often self-appointed or sponsored by outside aid agencies or international NGOs (National Research Council, 1992; Guyer, 1994; Mazonde, 1996). While grassroots organizations and NGOs can be very positive forces in rural development and in holding elected bodies accountable, the development literature provides many cases where membership organizations have failed to sustain their development efforts, or have benefited only a privileged minority (see Fox and Butler, 1987).

Projects and policies often prescribe the establishment of local NGOs or committees to manage and use natural resources. In Cameroon, India and Uganda these committees are sometimes elected. At times in Cameroon, Mali, Uganda, Zimbabwe, Bolivia, Mexico and Nicaragua, they are organized around interest groups. In contrast, in Mali, Bolivia and some cases in India, user committees and groups may be self-constituting and must present themselves to the elected local authorities for recognition. After some of these committees are constituted, local elected authorities allocate management use and powers to the committees. This strategy both strengthens local elected authorities and gives them the role of balancing interests among users. In this case, as is appropriate when public resources are involved, the chain of accountability is from the committee to the elected local government and from the local government to the people.

Chiefs, headmen and *other so-called 'customary authorities'* are often targeted by central governments, donors and NGOs as appropriate local authorities in decentralization efforts. Central government in Burkina Faso, South Africa, Uganda and Zimbabwe are reviving these authorities as the recipients of decentralized powers (Ribot, 1999a; Muhereza, 2001; Bazaara, 2002b; Cousins and Kepe, 2002; Ntsebeza, 2002). Some traditional authorities are very accountable to their people (Spierenburg, 1995). Many, however, are not (van Rouveroy et al, 1999; Kassibo, 2002a; Thiaw and Ribot, 2003; Ntsebeza, 2004). They often inherit their positions, and their degree of local accountability depends upon their personalities and local social and political histories. They may or may not be accountable to local populations. Although they are often depicted as legitimate, their legitimacy may be as much a product of fear as of respect (Ntsebeza, 2002, 2004), or may come entirely from powers and backing given to them by central government or donors.

Rather than enfranchising local people under democratic decentralization, choosing non-democratic authorities may – as under the colonial policies of 'indirect rule' and 'association' – subject local people to arbitrary authority without representation, rights or recourse (Mamdani, 1996; Ribot, 1999a). Customary authorities are notorious for entrenched gender inequalities and for favouring divisive ethnic-based membership – rather than the residency-based forms of citizenship so fundamental to most democratic systems (Vijayalakshmi, 2002). Today there is a troubling convergence of state and donor efforts to find the 'real', 'traditional' natural-resource managers and to empower them to manage the resources. But merely giving powers to customary authorities does not strengthen democratic decentralization.

Committees, NGOs and customary authorities – whether membership based, self-appointed, elected or oligarchic – are not necessarily accountable to elected local government. Selecting such semi-private or alternative representative bodies instead of giving public decision-making powers to elected bodies diminishes the role and authority of elected local government. Whether local authorities are elected or hereditary, they appear to be accountable to the donors and ministries that are transferring management roles and powers to them, rather than to the local people. Their accountability follows the purse strings (see Mamdani, 1996). They also often lack discretion due to a limited set of roles imposed from above. Due to these constraints, they often function as implementing agents for central authorities, rather than as local independent discretionary decision-makers. Although these institutions have many positive roles, they do not represent the public. Institutional plurality is important; but, unmediated by representation, it may serve only the best organized and most powerful interests and elites (Ribot, 2002a).

Many countries lack viable representative local governments. Accountability measures, with or without representative local government, can foster a degree of downward accountability of whichever authorities hold powers over the environment. Elections are not an exclusive means of achieving such accountability. Where elected local governments exist, working with them can be a first step towards supporting local democracy. Strengthening them is a second step. Where they do not exist at all, insisting that they be established is a priority. Reinforcing multiple forms of accountability for existing local institutions is the next best option (see Mandondo, 2000; Ribot, 2002a; 2004). In addition, attention must also be paid to making administrative bodies and other levels of government accountable to local governments so that they can deliver the services that local people expect and demand.

Discretionary powers

In most decentralizations, few discretionary powers over natural resources are transferred to local authorities. Yet, for effective decentralization to take place, meaningful discretionary power transfers are critical. Without discretionary powers, even the most accountable democratic local authorities can be irrelevant. Discretionary powers enable local authorities to respond flexibly to local needs and aspirations, making them relevant to their constituents. Discretionary powers also give local people a reason to engage with the state and to begin demanding that decisions conform to their needs. Civil society begins to organize and crystallize around empowered representative authorities (Anu Joshi, 1999, pers comm). There is no reason to organize and lobby representatives who hold no meaningful powers because they cannot be held accountable. Whereas power transfers without accountable representation can be dangerous, representation without powers is empty.

In current decentralization initiatives, many powers remain centralized that could be devolved to local authorities without threat to the environment (Fairhead and Leach, 1996; Ribot, 1999b; Conyers, 2001; Goldman, 2001). Forestry and wildlife agencies transfer use rights with no commercial value while retaining central control over the lucrative aspects of the sector (Ribot, 2001b, 2002a; Bazaara, 2002a). Management requirements are set by central governments that far exceed necessary measures (Fairhead and Leach, 1996; Leach and Mearns, 1996; Ribot, 1999a). Forest agencies commonly establish complex prescriptive systems of forest management planning, requiring 'expert' forestry agent approval before local governments can make decisions as to how, when or where forests are used and commercialized.

The most commonly transferred positive powers are tax and fee revenues from local natural resources. These revenues have made significant contributions to local communities to build schools, grain mills and other public projects. The power to allocate small concessions has been transferred to local governments in Bolivia, Cameroon, Indonesia, Mali and Zimbabwe; these rights have increased the power, and likely the legitimacy, of the local authorities. Although this represents a great advance in decentralizations, the right to revenues and the ability to allocate concessions or production permits is restricted in all of these cases to a small portion of the forested area under exploitation. The rest is reserved for the central government to exploit. Often, this right is also shrouded in overly extensive planning requirements and tight oversight, unduly restricting local discretion.

Research has identified several systematic problems in the selection and balance of powers being transferred to local authorities. Obligations and instrumental objectives of the central state are being transferred to local authorities in lieu of discretionary power. These mandates are often unfunded. Tax and fee revenues, when available, do not always cover the cost of obligations. Uses with no commercial value are transferred instead of lucrative opportunities. Technical decisions about management techniques are often conflated with non-technical decisions concerning allocation; the non-technical decisions are being retained centrally by the forest service, though they are precisely the kinds of decisions that could be made by local authorities with little threat to the resource (Bazaara, 2002b).

Geographic scale also affects the distribution of powers. Based on work in India, Agrawal (2001) points out that decentralization success can be fettered by jurisdictions being too large or even too small. For purposes of accountability, representation and participation, some political or administrative jurisdictions may be too large to be considered local, which is the case for the lowest level of local government in Burkina Faso. Often the relevant question is which scale is most appropriate for which kinds of decisions. In practice, matching jurisdictions with ecological formations cannot always be accomplished because watersheds and forests may not fall within a single local political or administrative jurisdiction. One approach to multiple geographic scales is to encourage the formation of local government federations and networks so that upstream and downstream constituents can work together for mutual benefits. This may be a better option than establishing new special-purpose districts (Ostrom, Schroeder, and Wynn, 1993).

The impulse to recentralize into higher-scale districts makes sense for some resources, but it must be cautiously approached. Adding more scales – or layers – of governance can take powers away from the local arena and concentrate them at higher levels (see Mandondo and Mapedza, 2003). The proliferation of scales being advocated by polycentric-governance proponents may diffuse powers among too many actors, rather than creating a cogent management system. The formation of federations and networks may be an effective approach to keeping governance local while attending to multi-scale problems. During recent years, the landscape approach has been introduced in natural resource management, moving decision-making powers to higher and higher scales that are geographically and technically designed and managed, rather than placing decision-making in the hands of residents of these landscapes (see for example, Central Africa Regional Programme for the Environment at www.carpe.umd.edu). These so-called landscape approaches must be explored for how they take power away from both the local populations and national governments while placing them in the hands of technical agencies and donor communities who claim to know the best manner in which to manage the resources.

Capacity arguments are consistently used by central ministries to block the transfer of powers to local authorities. 'Capacity' is a chicken-and-egg problem. There is reluctance on the part of central governments to devolve powers before capacity has been demonstrated; but without powers there is no basis upon which local authorities can gain the experience needed to build capacity. Nor is there any basis for demonstrating that capacity has been gained. Furthermore, arguments based on the lack of capacity are often used as excuses, rather than justified reasons, for not devolving powers (Fairhead and Leach, 1996; Brown, 1999; Ribot, 1999a; Conyers 2001). Strategies are needed so that powers can be transferred before capacity is demonstrated so that local empowerment has a chance of occurring.

The mix of powers and obligations to be retained at the centre and those to be devolved to lower political-administrative scales is a matter that requires critical analysis and informed public debate. Otherwise, environmental agencies are likely to continue to retain powers and micromanage environmental sectors, whether such measures are necessary or not. The principle of 'subsidiarity' calls for decisions to be made at the lowest possible political-administrative level (Føllesdal, 1998; Rocher and Rouillard, 1998); but this principle is not followed in most environmental decentralizations.

Principles to guide the division of executive, legislative and judiciary powers could include the following:

- Discretionary powers must be transferred to give local authorities some independence.
- These powers must have value or significance to local people.

- Mandates must be matched by sufficient fiscal resources and technical support, and mandates should not be the only powers transferred to local authorities.
- Commercially valuable resource-use opportunities should be transferred to local authorities in addition to subsistence-oriented usufruct rights.
- Technical decisions, some of which need to be made at a central level, must not be conflated with political decisions concerning use of resources (that is, who should have access to and benefit from them).
- Attention should be paid to the separation and balance of powers at each level of government.
- *Public* resources – including most forests, fisheries and pastures – should be kept within the public sector, not privatized.

Secure power transfer

Means of transfer is another critical dimension of decentralization (Conyers, 1990; Ahwoi, 2000). Security and sustainability of decentralization reforms rest largely on the means used to transfer powers from central government to other entities. Means of transfer can be constitutional, legislative, or may be accomplished through ministerial decrees or administrative orders. *Constitutional transfers* are the most secure and sustainable (see Conyers, 2000).

Ethiopia, Ghana, Mali, Senegal, South Africa and Uganda all have constitutional clauses that ensure some degree of government decentralization (UNCDF, 2000, p6). While these clauses do not specify which powers are decentralized, they provide leverage for law-makers to establish and maintain decentralized governance arrangements. The specification of the powers to be decentralized usually takes place through decrees and orders, which can change with the balance of powers among parties or with the whims of the party or administrators in power (Ahwoi, 2000). These powers are delegated, not securely transferred. As Oyugi (2000, p7) suggests, 'those receiving delegated authority act for those who delegate it'.

In environmental legislation in Burkina Faso, Cameroon, Guinea, Mali, Senegal, South Africa, Zimbabwe and elsewhere, decisions concerning the allocation of important powers are made by ministerial or administrative decree. In Mali, for example, decentralization is called for by the constitution, and decentralization of powers over natural resources is called for in environmental legislation, such as the 1996 forestry code. But within the forestry code, the powers to be devolved are specified by decree of the minister responsible for forests. Like many environmental agencies around the world, Mali's environmental service has yet to officially transfer any powers to local authorities. The decentralization in Mali's environmental sector is a discretionary matter for the ministry responsible for forests and its administrative staff. In this manner, what appears to be a constitutional guarantee is transformed into executive-branch discretion (Ribot, 2002a).

Until people believe that the rights they have gained are secure, they are not likely to invest in them. New rights to exploit forests may be exercised with urgency by people who believe that the government will take these rights away in the near future. In this manner, decentralization reforms may cause over-cutting. Such overexploitation following decentralization reforms was observed in Kumaon, India, during the 1930s before the situation stabilized. Central authorities can also subject local people to their will by threatening to withdraw powers, or can even transform elected local authorities from instruments of enfranchisement into central administrative agents.

As with land tenure security, the security of transfer of decision-making powers from central government to local institutions shapes the sustainability of the reforms and the willingness of local people to believe and invest in the reforms. Transfers made by legislative reforms are more secure than those made by ministerial decrees, administrative orders or the discretion of administrative authorities. Insecure means of transfer discourage local people from investing in new decentralization laws. The environmental ramification of this phenomenon is that people are more likely to over-exploit resources while they can, and are less likely to invest in environmental maintenance if they do not believe their new privileges will last. In Indonesia, over-exploitation of forests may be due to this kind of dynamic (Resosudarmo, 2002). Retaining inordinate discretion in the executive branch also creates opportunities for allocation along political and social lines to serve the interests of central agents. Secure means of transfer may help to reduce such opportunities for abuse and corruption.

Central government roles

Decentralization can benefit from a strong central state; in fact, as Conyers (2000, p22) points out:

> *Ironically, decentralization policies are most likely to be implemented effectively in situations where the government is politically secure and power is concentrated in the hands of a relatively small group of people. A secure government can afford to decentralize a substantial amount of power without threatening its own existence.*

Paradoxically, structural adjustment programmes that promote decentralizations at the centre often appear to undermine the establishment of sound local government by depriving central governments of the funds and staff that are needed to support successful local reforms (Crook and Sverrisson, 2001). Many powers belong with central government, such as establishing the legal enabling environment for decentralization, setting national environmental priorities and standards, establishing poverty-reduction strategies and ensuring compliance with national laws. Central government also has roles in supporting a variety of local efforts with finance and technical services.

Minimum environmental standards: A necessary and logical complement to decentralization

Even perfectly representative and downwardly accountable local authorities may over-exploit resources and ignore minority interests if given the unbridled power to do so. When it is profitable, decision-makers are likely to exploit natural resources rather than conserve them, especially if they do not bear the indirect costs, such as downstream flooding and dam siltation from upstream deforestation. When the present needs are especially urgent, and local costs of exploitation are not immediately incurred, resources are exploited. Achieving positive environmental and social outcomes requires standards and means for ensuring that nationally defined environmental and social concerns are taken into account. Of course, outcomes also depend upon the local history and culture of conservation, cooperation and social stratification.

Foresters and environmentalists often complain that local people cannot make all natural resources management decisions. No reasonable decentralization advocates, however, are calling for the transfer of *all* decisions over natural resources to local populations. Subsidiarity principles are one means of determining which powers can be transferred to local people without threatening the integrity of natural resources or social well-being; minimum environmental standards are a complementary means of codifying these principles in law, thus establishing greater local autonomy in natural resources management and use. The minimum-standards approach complements decentralization by specifying the boundaries to the domain of local autonomy without restricting discretion within those boundaries – for example, without requiring pre-approval by government agents.

A minimum environmental standards approach – a set of restrictions and guidelines for environmental use and management – would replace the centrally directed micromanagement approach currently exercised through elaborate plans and planning processes. Local representative authorities must enforce these standards, make public management and use decisions and mediate disputes among users. Some kinds of actions may require management plans in order to maintain the minimum standards; but permission is not required from central environmental ministries unless activities violate or require modification of the minimum standards.

Establishing minimum standards is an important role of central governments. But it must be done based on sound social and ecological research within an open political process following clear subsidiarity principles so that central government cannot, as it may want to, retain unnecessary control over forest management. Central government does not want to give up control over resources any more than the fox wants to give up access to the hen house.

Uniform minimum standards

In forest management, different rules and requirements are often applied to different actors. Communities are often required under many national laws to manage forests more rigorously than are corporations. The entry barriers posed by multiple requirements on local communities can prevent communities from entering into forest management. Uniform minimum standards – that require the same thing of communities as of corporations – can help to avoid double standards that exclude local communities (see Ribot, 2004).

Additional measures for equity

Decentralization can shape equity within and among local districts. Inter-jurisdictional equity depends upon the government's willingness to engage in redistribution of resources among districts. Decentralization can result in a situation where localities endowed with good natural, financial or technical resources prosper at the expense of those without (Conyers, 2000, p8). Remedies to inter-jurisdictional inequalities require the central state to engage in redistribution (Smoke, 1999; World Bank, 2000).

While poverty alleviation is often assumed to be one of the positive outcomes of decentralized governance, a comparative study of decentralization and poverty alleviation concludes that, within districts, 'responsiveness to the poor is quite a rare outcome' and 'positive outcomes are mainly associated with strong commitment by a national government or party to promoting the interests of the poor at the local level' (Crook and Sverrisson, 2001). Central governments tend to be more generous towards the poor than local governments. In decentralizations concerning natural resources, inequitable local decision-making and benefit distribution are frequently observed. Local elites may be more prejudiced against the poor than those at higher levels. Dominant ethnic groups can use their new powers to take advantage of weaker ones (Crook and Sverrisson, 2001; James Manor, cited in Latif, 2002).

Local mediation mechanisms

In redistributing rights over resources, decentralization produces winners and losers. Conflicts emerge as decision-making processes change, and tensions arise over the use of natural resource revenues. Conflicts also emerge among users, local authorities and governmental natural resources management agencies. In most cases, conflicts over natural resources are adjudicated by the government agency responsible for the resource in question. But this arrangement creates conflicts of interest and unfair outcomes due to the failure to separate executive from judiciary functions.

Local dispute resolution mechanisms, accessible courts and channels of appeal outside of the central government agencies involved are needed to facilitate a smooth transition to decentralized systems of environmental governance. Setting up official adjudication systems is the responsibility of central government. Alternative dispute resolution mechanisms designed to supplement, but not replace, a fair judiciary can also be enabled by central government and supported by communities, donors and NGOs.

Civic and local government education

Civic education can ensure that people know their rights, as well as the obligations that government has towards them. In turn, education can ensure that local governments know their powers and understand their obligations to local people. Informing people of their rights, writing new laws in clear and accessible language and translating new legal texts into local languages can encourage citizen engagement and local government responsibility. Central government, donors and NGOs can support such local efforts.

ASKING THE RIGHT QUESTIONS

Practitioners, donors, policy-makers and activists need to ask straightforward questions in order to evaluate whether the institutions and powers being chosen in decentralization reforms are likely to provide the equity, efficiency, development and environment benefits that decentralization promises. These questions are designed to help guide design and implementation of decentralization reforms. The following questions are taken from Ribot (2004).

Institutional choice questions

What kind of institution is receiving powers in the name of decentralization? Are they elected local government or local administrative authorities; local branches of line ministries or traditional authorities; NGOs, associations, appointed committees or elected committees?

If the local institutions are elected, do the electoral rules help to make them representative? How long in advance are elections announced? Is there universal suffrage? Is it residency based? How are candidates chosen? Do electoral laws admit independent candidates? How long are term lengths? Are there means of recall?

To whom is the local institution accountable with respect to the exercise of the transferred powers? Through what mechanisms is the local institution accountable? Are there multiple mechanisms of accountability?

How does the origin of their funding or their powers affect their accountability? Can they raise revenues locally? Do they depend upon grants and funding from outside agents?

How does the mechanism through which resources are transferred affect their accountability? Are the powers they receive transferred as secure rights? Are the powers transferred as privileges that can be taken away?

Is the institution integrative across sectors? Is the institution multi- or single sector? Is it multi- or single purpose? Does its role include mediating among sectors? Does its role include allocation of resources among sectors?

Does the institution favour procedural matters of democracy or the specific set of instrumental objectives?

Is the form of inclusion, belonging or citizenship based upon residency, identity or interest?

Is the institution favourable towards marginal and poor populations? Do marginal and poor populations have influence over and voice in the institution? Are there mechanisms to ensure the inclusion of women and of other marginal or poor populations in decision-making and benefits?

Whose interests are ultimately served by the chosen local institution(s)? Is the institution serving patronage interests of central actors? Is the institution serving only the interests of its members? Is the institution serving only a sub-sector or fraction of the population? Is the institution serving the population as a whole?

Can multiple institutions freely function in the local arena? Do citizens have rights to organize? Do citizens and local organizations have rights to lobby government? Can groups easily attain legal recognition and status?

Are lines of accountability over public decisions mediated through representative authorities? When non-representative institutions are given public decision-making powers, are they accountable to representative authorities concerning the exercise of these powers? Do these institutions compete with and undermine representative authorities, or do they strengthen representative authorities?

What are the long-term implications of the choice of institutions for justice, sustainability, scaling up and the formation of citizenship? Do these institutions encourage broad-based involvement of local people? Do they enfranchise people as citizens? Do they give local people voice and agency? Do they enable long-term stability? Are they replicable across territories?

Power choice questions

Does the transfer of power result in discretion for local decision-makers?

Are the powers being transferred meaningful to local people?

Are the powers being transferred significant enough to engage local people with local decision-makers?

Are mandates (obligations) being transferred?

Are those mandates sufficiently funded?

Are those mandates within the capacity of local authorities to implement?

Are political choices – such as who can use a resource and who can benefit – being retained at the centre or transferred to local decision-makers?

Are resources (lands, pastures, forests, fisheries) that have been accessible to the public and serve the public interest being privatized?

Are transfers of power made in a secure manner or can they be taken away at the whim of central authorities?

Is there appropriate separation of executive and legislative powers in the local arena and within agencies of central government?

Is there sufficient power – executive, legislative, judicial – in the local arena to balance and fight central interests?

If exploitation is allowed, are local authorities receiving the right to determine who exploits the resource? Can they decide over and mediate who has subsistence access to the resource? Can they decide who can exploit the resource commercially – that is, can they allocate exploitation rights?

Do local authorities have the right to say 'no' to commercial exploitation of local resources – that is, do they have the right to conserve the resource?

Are the powers that are transferred well matched to the political-administrative scale and to the ecological scale of the resource?

Is the environmental service using a system that requires approval for every decision, or are there decisions that can be made locally under an environmental standards framework?

Do environmental laws treat communities differently than commercial interests in a manner that excludes local communities from decision-making and benefits?

Are the skills required before transfers take place really necessary from an ecological perspective, or can decisions be transferred prior to demonstrating capacity?

Conclusion

The potential of decentralization to be efficient and equitable depends upon the representativeness of local institutions. But there are few cases where democratically accountable local institutions are being chosen *and* given discretionary powers. Before decentralizations can be evaluated, time is needed for them to be legislated and implemented and to take effect. First, locally accountable representation with discretionary power must be established. Then, accompanying measures must be identified to ensure environmental protection, justice and freedom from conflict. To encourage the decentralization experiment and test the conditions under which it yields the promised benefits, decentralization must be tested, monitored and evaluated. Research will be needed to determine:

- whether decentralizations are being established; and
- their social and environmental effects.

To begin with, practitioners, donors, activists, policy-makers and researchers can ask questions to determine whether a decentralization effort is serious and is likely to result in the conditions that theory tells us will produce positive outcomes.

Note

The contents of this chapter are condensed from two published articles (Ribot 2002b, 2004). The material in these articles could not have been compiled without the help of colleagues around the world. Many of these colleagues are cited throughout the chapter. I owe great thanks for case material provided by the members of the World Resources Institute's (WRI's) research project on Accountability, Decentralization and Environment in Africa: Nyangabyaki Bazaara, Patrice Bigombé Logo, Diana Conyers, Thierno Diallo, Patrice Etoungou, Juliet Kanyesigye, Bréhima Kassibo, Naffet Keita, Alois Mandondo, Everisto Mapedza, Frank Emmanuel Muhereza, Eugene Muramira, Agrippinah Namara, Xavier Nsabagasani, Lungisile Ntsebeza, Phil René Oyono and Alda Salomao. I am also grateful to the participants in the Bellagio Workshop on Decentralization and the Environment, whose research and observations all contributed to this chapter. In addition to Bazaara, Kassibo, Ntsebeza, and Oyono, mentioned above, they include Arun Agrawal, Amita Baviskar, Christian Brannstrom, Ben Cousins, Mairi Dupar, Xu Jianchu, David Kaimowitz, Anne Larson, Mehr Latif, James Manor, Robin Mearns, Fernando Melo Farrera, James Murombedzi, Pablo Pacheco, Nancy Peluso, Ida Aju Pradnja Resosudarmo, Roberto Sanchez and Uraivan Tan-Kim-Yong. I want to thank the Center for International Forestry Research (CIFOR) in Indonesia and Cameroon, and the Centre for Basic Research in Uganda for partnering in these efforts. I want to give special thanks to Jon Anderson of the US Agency for International Development's Economic Growth, Agriculture and Technology division for supporting field-based policy research. Rockefeller's Bellagio Center, the Ford Foundation, the MacArthur Foundation, the Dutch government and the Africa Bureau of the US Agency for International Development (USAID) have also contributed substantially to these efforts. Most of all, I want to thank Peter G. Veit at the World Resources Institute for his collegial support throughout the decentralization and environment research programme. Many thanks to Carol Pierce Colfer of CIFOR and to two anonymous reviewers for their constructive comments.

References

Agrawal, A. (2001) 'The regulatory community: Decentralization and the environment in the *van panchayats* (forest councils) of Kumaon, India'. *Mountain Research and Development*, vol 21, pp208–211; paper also presented at the World Resources Institute Workshop on Decentralization and the Environment, Bellagio, Italy, 18–22 February 2002

Agrawal, A. and Ribot, J. C. (1999) 'Accountability in decentralization: A framework with South Asian and African cases'. *Journal of Developing Areas*, vol 33, pp473–502

Ahwoi, K. (2000) 'Challenges facing local governments in Africa in the 21st century: An executive summary'. *Local Government Perspectives*, vol 7, pp1–7

Baviskar, A. (2002) 'Between the micro-politics and administrative imperatives: Decentralization and the watershed mission in Madhya Pradesh, India'. Paper presented at the World Resources Institute Workshop on Decentralization and the Environment, Bellagio, Italy, 18–22 February

Baviskar, A. (2004) 'Between micro-politics and administrative imperatives: Decentralization and the watershed mission in Madhya Pradesh, India'. *European Journal of Development Research*, vol 16, pp26–40

Bazaara, N. (2002a) 'Actors, powers and environmental accountability in Uganda's decentralization'. Paper presented at the World Resources Institute Workshop on Decentralization and the Environment, Bellagio, Italy, 18–22 February

Bazaara, N. (2002b) 'Decentralisation, politics and environment in Uganda'. Draft environmental governance working paper, World Resources Institute, Washington, DC

Blair, H. (2000) 'Participation and accountability at the periphery: Democratic local

governance in six countries'. *World Development*, vol 28, pp21–39
Brown, D. (1999) 'Principles and practice of forest co-management: Evidence from West-Central Africa'. *European Union Tropical Forestry Paper 2*, Overseas Development Institute, London
Conyers, D. (1990) 'Decentralization and development planning: A comparative perspective', in de Valk, P. and Wekwete, K. H. (eds) *Decentralizing for Participatory Planning: Comparing the Experiences of Zimbabwe and Other Anglophone Countries in Eastern and Southern Africa*. Avebury Press, Aldershot, UK
Conyers, D. (2000) 'Decentralisation: A conceptual analysis, parts 1 and 2'. *Local Government Perspectives: News and Views on Local Government in Sub-Saharan Africa*, vol 7, nos 3, 4, pp7–9, 13 (in part 1) and 18–24 (in part 2)
Conyers, D. (2001) *Whose Elephants Are They? Decentralization of Control over Wildlife Management through the CAMPFIRE Programme in Binga District, Zimbabwe*. Working paper, 31 January, World Resources Institute, Washington, DC
Cousins, B. and Kepe, T. (2002) 'Decentralization when land and resources rights are deeply contested: A case study of the Mkambati eco-tourism project on the wild coast of South Africa'. Paper presented at the World Resources Institute Workshop on Decentralization and the Environment, Bellagio, Italy, 18–22 February
Crook, R. C. and Sverrisson, A. S. (2001) *Decentralization and Poverty Alleviation in Developing Countries: A Comparative Analysis* or, *Is West Bengal Unique?* Institute for Development Studies working paper 130, IDS, Brighton
Dupar, M. and Badenoch, N. (2002) *Environment, Livelihoods, and Local Institutions: Decentralization in Mainland Southeast Asia*. World Resources Institute, Washington, DC
Evans, P. B. (1997) 'The eclipse of the state? Reflections on stateness in an era of globalization'. *World Politics*, vol 50, pp62–87
Fairhead, J. and Leach, M. (1996) *Misreading the African Landscape: Society and Ecology in a Forest-Savanna Mosaic*. Cambridge University Press, Cambridge
Ferguson, J. (1996) *Transnational Topographies of Power: Beyond 'the State' and 'Civil Society' in the Study of African Politics*. Unpublished manuscript, mimeo
Føllesdal, A. (1998) 'Survey article: Subsidiarity'. *The Journal of Political Philosophy*, vol 6, pp190–218
Fox, J. and Butler, J. (1987) *The Study of Membership Organizations: A Progress Report*. Report to the Inter-America Foundation, Washington, DC, May 18
Guyer, J. (1994) 'The spatial dimensions of civil society in Africa: An anthropologist looks at Nigeria', in Harbeson, J. W., Rothchild, D. and Chazan, N. (eds) *Civil Society and the State in Africa*. Lynne Rienner, Boulder, CO
Kassibo, B. (2002a) 'Participatory management and democratic decentralization management of the Samori Forest in Babye commune, Mopti region, Mali'. Paper presented at the World Resources Institute Workshop on Decentralization and the Environment, Bellagio, Italy, 18–22 February
Kassibo, B. (2002b) *Fondements Historiques et Politiques de la Gestion Décentralisée des Ressources Naturelles au Mali – La Commune en Questions: Synthèse de Deux Études de Cas*. Draft environmental governance working paper, World Resources Institute, Washington, DC
Keita, N. (2002) *Décentralisation et Responsabilité dans les Modes de Gestion des Ressources Naturelles et des Redevances y Afférentes*. Les Bourgoutières Yallarbe de Youwarou, draft environmental governance working paper, World Resources Institute, Washington, DC
Larson, A. (2004) 'Formal decentralization and the imperative of decentralization "from below": A case study of natural resource management in Nicaragua'. *European*

Journal of Development Research, vol 16, no 4, pp55–70

Larson, A. and Ribot, J. C. (2005) 'Democratic decentralization through a natural resource lens: An introduction' in Ribot, J. C. and Larson, A. M. (eds) *Democratic Decentralisation through a Natural Resource Lens*. Routledge, London, pp1–25

Latif, M. (2002) Minutes of the World Resources Institute Workshop on Decentralization and the Environment. Bellagio, Italy, 18–22 February

Leach, M. and Mearns, R. (1996) *The Lie of the Land: Challenging Received Wisdom on the African Environment*. James Curry, Oxford

Mamdani, M. (1996) *Citizen and Subject: Contemporary Africa and the Legacy of Late Colonialism*. Princeton University Press, Princeton

Manin, B., Przeworski, A. and Stokes, S. (1999) 'Elections and representation', in Manin, B., Przeworski, A. and Stokes, S. (eds) *Democracy, Accountability, and Representation*. Cambridge University Press, Cambridge

Mandondo, A. (2000) *Situating Zimbabwe's Natural Resource Governance Systems in History*. Occasional Paper No 31, December 2000, CIFOR, Bogor, Indonesia

Mandondo, A. and Mapedza, E. (2003) *Allocation of Government Authority and Responsibility in Tiered Governance Systems: The Case of Environment-related Laws in Zimbabwe*. Environmental Governance in Africa Working Paper, World Resources Institute, Washington, DC

Mazonde, I. N. (1996) 'The Basarwa of Botswana: Leadership, legitimacy and participation in development sites'. *Cultural Survival Quarterly*, vol 20, www.culturalsurvival.org/publications/csq/csq_article.cfm?id=24FDDF3F-82C8-4364-A91E-066FA84F6DDC&region_id=1&subregion_id=11&issue_id=24

Melo Farrera, F. (2002) 'Local diversity to contend with globalization in Oaxaca, Mexico. Paper presented at the World Resources Institute Workshop on Decentralization and the Environment, Bellagio, Italy, 18–22 February

Moore, M. (1998) 'Death without taxes: Democracy, state capacity and aid dependence in the Fourth World'. In White, B. and Robinson, M. (eds) *Towards a Democratic Developmental State*, Oxford University Press, Oxford

Muhereza, F. (2001) 'Concept note on the need to re-think the issue of private forests'. February 2001, Centre for Basic research, Kampala

National Research Council (1992) *Democratization in Africa: African Views, African Voices*, Sahr John Kpundeh (ed). National Academy Press, Washington, DC

Ntsebeza, L. (2002) 'Decentralization and natural resource management in rural South Africa: Problems and prospects'. Paper presented at the World Resources Institute Workshop on Decentralization and the Environment, Bellagio, Italy, 18–22 February

Ntsebeza, L. (2004) 'Democratic decentralization and traditional authority: Dilemmas of land administration in rural South Africa'. *European Journal of Development Research*, vol 16, pp71–89

Oyugi, W. O. (2000) 'Decentralization for good governance and development'. *Regional Development Dialogue*, vol 21, pp3–22

Pacheco, P. (2004) 'What lies behind decentralization? Forest, powers and actors in lowland Bolivia'. *European Journal of Development Research*, vol 16, pp90–109

Resosudarmo, I. A. P. (2002) 'Closer to people and trees: Will decentralization work for the people and the forests of Indonesia?' Paper presented at the World Resources Institute Workshop on Decentralization and the Environment, Bellagio, Italy, 18–22 February

Resosudarmo, I. A. P. (2002) 'Closer to people and trees: Will decentralization work for the people and the forests of Indonesia?' Paper presented at the WRI's Conference on Decentralization and the Environment, Bellagio, Italy, 18–22 February

Resosudarmo, I. A. P. (2002) 'Closer to people and trees: Will decentralization work for the people and the forests of Indonesia?' In Ribot, J. C. and Larson, A. M. (eds)

Decentralization of Natural Resources: Experiences in Africa, Asia and Latin America, Frank Cass, London
Ribot, J. C. (1999a) *Integral Local Development: Authority, Accountability, and Entrustment in Natural Resource Management.* Working paper of the Africa Region RPTES Programme, April, World Bank, Washington, DC
Ribot, J. C. (1999a) 'Decentralization and participation in Sahelian forestry: Legal instruments of central political-administrative control'. *Africa*, vol 69, no 1, pp23–65
Ribot, J. C. (1999b) 'A history of fear: Imagining deforestation in the West African Sahel'. *Global Ecology and Biogeography*, vol 8, pp291–300
Ribot, J. C. (2001a) 'Integral local development: Accommodating multiple interests through entrustment and accountable representation'. *International Journal of Agricultural Resources, Governance and Ecology*, vol 1, pp327–350
Ribot, J. C. (2001b) *Science, Use Rights, and Exclusion: A History of Forestry in Francophone West Africa.* Issue paper no 104, Drylands Programme, International Institute for Environment and Development (IIED), London
Ribot, J. C. (2002a) *African Decentralization: Local Actors, Powers and Accountability.* United Nations Research Institute on Social Development (UNRISD) Programme on Democracy, Governance, and Human Rights, paper no 8, UNRISD, Geneva
Ribot, J. C. (2002b) *Democratic Decentralization of Natural Resources: Institutionalizing Popular Participation.* World Resources Institute, Washington, DC
Ribot, J. C. (2004) *Decentralization of Natural Resource Management: Encountering and Countering Resistance.* World Resources Institute, Washington, DC
Rocher, F. and Rouillard, C. (1998) 'Decentralization, subsidiarity, and neoliberalism in Canada: When the tree hides the forest'. *Canadian Public Policy-Analyse de Politiques*, vol 24, pp233–258
Romeo, L. (1996) *Local Development Funds: Promoting Decentralized Planning and Financing of Rural Development.* Policy series, United Nations Capital Development Fund, New York
Schroeder, R. A. (1999) 'Community, forestry and conditionality in the Gambia'. *Africa*, vol 69, pp1–22
Smoke, P. (1999) 'Understanding decentralization in Asia: An overview of key issues and challenges'. *Regional Development Dialogue*, vol 20, pp1–17
Spierenburg, M. (1995) 'The role of the Mhondoro cult in the struggle for control over land in Dande (northern Zimbabwe): Social commentaries and the influence of adherents'. Center for Applied Social Sciences, University of Zimbabwe, Harare, and Amsterdam School for Social Science Research, Amsterdam
Tendler, J. (1997) *Good Government in the Tropics.* Johns Hopkins University Press, Baltimore
Thiaw, S. and Ribot, J. C. (2003) 'Insiders out: Forest access through villager chiefs in Senegal'. Paper presented at the International Conference on Competing Jurisdictions: Settling Land Claims in Africa, Vrije Universiteit, Amsterdam, 24–27 September
UNCDF (United Nations Capital Development Fund) (2000) *Africa: Decentralisation and Local Governance Conference Concept Paper.* Draft paper for the Decentralisation and Local Governance Conference, UNCDF, Cape Town, March 2001
van Rouveroy, van Nieuwaal, A. B. E. and van Dijk, R. (eds) (1999) *African Chieftaincy in a New Socio-political Landscape.* LIT Verlag, Hamburg
Vijayalakshmi, V. (2002) Gender Accountability and Political Representation in Local Government. Working paper no 102, Institute for Social and Economic Change, Bangalore
World Bank (2000) *Entering the 21st Century: World Development Report 1999/2000.* Oxford University Press, Oxford

Chapter 5

The Experience of the Food and Agriculture Organization with Decentralization in the Forest Sector

Merilio G. Morell

Introduction

In this chapter, the United Nations Food and Agriculture Organization's (FAO's) experience with decentralization is briefly reviewed. Two countries – Burkina Faso and Mali – that are implementing their decentralization laws with the FAO's technical assistance provide examples. The required expertise, intervention opportunities and strategies used to implement decentralized governmental administration of forest resources are analysed. Conclusions and recommendations are formulated on the conditions and factors needed to ensure that decentralized administration of forest resources results in improved sustainability, people's participation, government accountability and equity in the sharing of benefits from forest resources. Information is also provided about the process of international technical assistance, how to assist countries in translating laws and regulations into concrete new institutional arrangements, and pointers for future work in this area.

Decentralization and achieving global development goals

Decentralization is not an end in itself; it is one of several means being used to achieve priority global development goals (UNDP, 2003). Fighting poverty, hunger and environmental deterioration are just a few of the goals on the world agenda. The intellectual approaches of the leading international development institutions and donors – the World Bank, the International Monetary Fund (IMF), the United Nations (UN), the Organisation for Economic Co-operation and Development, the European Union (EU), the Group of 8 (G8) and national governments – converge in considering 'quality economic growth' the overall necessary condition for achieving these goals (Camdessus, 1999).

For quality growth to occur, country governments must be capable of ensuring competitive participation in free trade, economic efficiency and equity and macro-economic stability – in other words, they must be governments capable of practising good governance. In turn, it is considered that good governance starts at the level of communities, people and local government participation. Decentralization is the means of allowing the participation of people and local governments. Securing good governance is a precondition to the quality growth necessary to achieve global development goals. This connection explains the increased efforts made to promote and support the implementation of the decentralization process in developing countries since the early 1980s. According to Ayres (undated), more than 80 per cent of the developing and transition countries of Eastern and Central Europe and the former Soviet Union, despite their widely different political orientations and economic bases, are experimenting with decentralization. Decentralization processes have focused first on the political, and then on the fiscal and administrative aspects. The rationale as well as the arguments for and against decentralization and strategies to improve the level of development of countries have been extensively discussed and analysed (World Bank, 1984, 1998, 2000a, b; Rondinelli et al, 1984; IDB, 1997; Manor, 1997; Ter-Minassian, 1997; Litvack et al, 1998 . They are therefore not dealt with at length in this chapter.

Political, fiscal and administrative decentralization exercises affect the forest sector and, in many cases, have raised the need for taking actions that otherwise would never have been considered or would have been implemented more slowly and on a smaller scale. These exercises are consequently the driving force for the current decentralization trends in the forestry sector.

Initial processes to implement decentralization in the forest sector have resulted in requests for technical assistance from the FAO's forestry department in decentralizing forest and natural resources management. This chapter reviews the experience of the forestry department in implementing decentralization in Burkina Faso and Mali, summarizing some of the activities undertaken and the main issues and challenges that arise from the experience.

FAO and Decentralization

The FAO has recognized the need to support its members with the decentralization process. It also recognizes that 'decentralization is a complex process requiring some enabling conditions to be sustainable, especially a strong continuing education programme for personnel, ordinary citizens and organizations that are assigned to decentralization functions and responsibilities' (Mai, 1999, p33). In the forestry sector, the FAO – with the support of government members and through the forestry department – has pioneered work in community and participatory forestry. The Forests, Trees and People Programme began in 1987 as a follow-up to the FAO's first focused programme to promote community forestry – the FAO/Swedish International Development Cooperation Agency (Sida) Forestry for Local Community Development Programme. This work anticipated many of the principles and premises that sustain decentralization in actual practice and allowed the FAO to gain first-hand experience in programme decentralization and institutionalization. The FAO has also worked on matters relating to agricultural policy and decentralized rural development. This work has been carried out under the responsibility of the Policy Assistance Division of the FAO's Department of Technical Cooperation. The division concentrates on three areas: public institutions, civil society and local government.

Community and participatory forestry

During the 1970s the FAO joined social scientists and foresters and started to explore how forestry could be used as a resource and tool of rural development in 'community-focused forestry'. Following this initial goal, a programme was initiated that ran without interruption from the late 1970s until 2002. The FAO's programme in community-participatory forestry can be divided into five phases. Phases I and II (1978–1986) were known as Forestry for Community Development. The goals of Phase I were to explore what community forestry could offer and why it might be important, and to promote awareness of this new potential. A major audience was forest policy-makers. Phase II, in addition to dealing with the *what* and *why* of community forestry, was intended to respond to the *how* by developing tools, methods and approaches for implementing community forestry. Phase II continued to involve forestry policy-makers in an advisory capacity, but also provided support to field-level foresters. Phase III, called Forests, Trees and People, developed between 1987 and 1991, focused on the development, publication and distribution of reports and material in eight countries. The overall intent was to strengthen the in-house platform for promoting and backstopping community forestry.

Phase IV of Forests, Trees and People began in 1992 and, after several extensions, concluded in 2002. Its overall action can be characterized as programme decentralization. The goal was to ensure that learning about community forestry and its application should be based on grassroots input, involvement and needs. Its intent was to decentralize the efforts to put community forestry into action and to identify ways of institutionalizing community forestry and participatory approaches.

Through its work in community and participatory forestry, the FAO gained experience in implementing the decentralization of technical assistance programmes. Another important link between that work and current decentralization is the many experiences and tools for improving people's participation in the planning, decision-making and management of natural resources.

Public institutions, civil society and local government

In December 1997, the FAO hosted a Technical Consultation on Decentralization in collaboration with the United Nations Capital Development Fund (UNCD), the International Fund for Agricultural Development (IFAD), the German Agency for Technical Cooperation (GTZ), the Swiss Agency for Development and Cooperation (SDC) and the World Bank. This consultation had two objectives: first, to draw lessons from a review of research and empirical evidence and to reach consensus on the potential and limitations of decentralization for rural development; and, second, to identify good practices and principles for the sequencing and design of decentralization strategies.

The consultation led to the preparation of 18 papers highlighting the capacities and knowledge accumulated by the FAO on decentralization and rural development. Although the focus of this consultation was on rural development, the analyses are relevant to decentralization in the forest sector.

Decentralization in Mali and Burkina Faso

In Burkina Faso and Mali, the FAO is helping to implement decentralization laws. Work is being carried out on responsibilities and functions, the extent and type of resources transferred, and the relationship with the decentralization of other sectors. The final result of this assistance will be designing and launching systems for the decentralized governance of forest resources.

Burkina Faso

From 1983 to 1987, the country adopted reforms that resulted in a certain degree of territorial and administrative decentralization. More definite action started in 1990 during the preparation of the draft for the constitutional law. The constitution, approved in 1991, established a territorial organization based on municipalities (*collectivités locales*). The municipalities are self-managed units under the guidance of councils elected at the local level. Since 1991 Burkina Faso has been taking measures for the structural adjustment of its economy. Market liberalization has resulted in the dismantling of some state enterprises, and a more important role has been assigned to the private sector and civil society. In this context, political and fiscal decentralization were launched.

In 1993 the laws regulating the territorial division of the country and the electoral procedure for municipal councils were enacted. By 1995, 33 municipalities were ready to participate in the local elections held that year. The next local elections were to be held in 2000; however, they were postponed until 2003, mostly because of a lack of institutional and human resources capacity at the municipal level.

In 1998 the laws establishing the general orientation for decentralization (objectives, territorial organization, principles for self-management of municipalities and the relationship between the state and the municipalities) were enacted; they also established the administrative regulations. The same year the law for implementing decentralization was promulgated. This law set forth the general guidelines for making decentralization a reality in Burkina Faso, with provisions for

- progressive implementation of decentralization;
- application of the principle of subsidiarity;
- joint transfer of responsibilities and of their associated resources; and
- creation of funds and special support for strengthening the capacity of local governments so that they could carry out their responsibilities.

Three organizations arose from these laws. The first was the National Commission for Decentralization (CND, from its French acronym), created in 1993 with the mandate to study the municipalities' financial situation, analyse the municipal–state division of responsibility, and provide capacity-building to local elected authorities and information management of the decentralization processes at national level. As a result of the work and recommendations of CND, two other organizations were set up. One was a support service for the management and development of municipalities (Service d'Appui à la Gestion et au Dévelopment des Communes, or SAGEDEDOM), responsible mainly for capacity-building for local authorities in the management of municipalities. The other organization was the fund for the development of municipalities (Fonds de Dévelopment des Communes, or FODECOM), whose main purpose is to provide financial aid to the municipalities for operational expenses and equipment acquisitions. Many other structures of smaller scope and level were created in the different ministries for accompanying the decentralization movement. One example is the Strategies and Methods Unit (Cellule Stratégie et Méthode, or CSM), created within the forestry department of the Ministry of Environment, whose main task is to study and plan decentralization in the forestry sector.

At the municipal level, decentralization laws have established only formal representation for the political party; there are no provisions for representation of civil society. Such an arrangement has obviously not considered the increased importance of civil society at the municipal and village level. The responsibilities of the communes cover the areas regarding land management and urbanism, economic development, planning, environment, management of natural resources, health, education and urban services. The new territorial organization that began in January 2004 was intended to create more than 300 municipalities.

Mali

The first steps towards decentralization in Mali go back to the 1960s with the laws on territorial organization. These laws created a territorial organization arranged as six layers of nested hierarchical units: region, circle, *arrondissement,* commune, village, and tribe or faction. Legal dispositions favouring strong local participation through the creation of local administrative organs were enacted in 1966. Similar to the forestry laws in many countries, the laws for territorial organization, while sophisticated in their rhetoric, remained mere text with little effect on the country's development.

The movement for a higher degree of decentralization took on new life in 1990 with the armed revolt in the north of the country. The main demand of the inhabitants of the north, mostly from the Tuareg ethnic group, was a higher degree of autonomy. In 1991, during a national conference with the participation of major local actors, the principles of administrative decentralization under a unitary state were developed. Enshrined in the 1992 constitution, decentralization was supposed to be oriented in two main directions:

1. the principle of free self-management of the municipalities through councils elected under conditions determined by the law; and
2. the creation of a high council of municipalities with the functions of a second chamber of the parliament, with special rights regarding decentralization and local development.

In 1993 the laws and decrees that form the legal framework for decentralization and its implementation were enacted.

At the institutional level the responsibilities for implementing decentralization were given to the Ministry of Territorial Administration through an administrative unit called Mission for Decentralization. After 1998 this responsibility was passed to the president of Mali. The first task of the mission of decentralization was to hold municipal elections and to set up the first group of elected authorities. The first municipal elections were held during 1998 and 1999. These elections marked the launching of political and administrative decentralization in Mali.

The final reform of the territorial administration increased the number of municipalities with right to self-management from 19 to 702. The levels of decentralization have also been increased from one to three (region, circle and municipality).

From 1993 to 2004, the World Bank and other organizations, through four projects, provided US$223 million to Mali in loans and assistance related to decentralization, including support for private-sector-based grassroots initiatives to alleviate poverty and management of natural resources. This sum does not, however, represent the total investment made in Mali to support decentralization during that period.

TECHNICAL ASSISTANCE FOR DECENTRALIZATION

The development of a governance system for Mali – a country that has 3 hierarchical levels, 702 municipalities, and the responsibilities of the local, central and traditional governments and authorities under a participative scheme – is the challenge facing administrative and fiscal decentralization, as well as forestry decentralization. Burkina Faso has a similar challenge in planning and implementing decentralized forest resources governance: the country has 13 regions, 45 provinces, more than 300 municipalities, 3 hierarchical levels and at least 3 types of authorities.

Scope

The FAO is assisting in implementing approved laws dealing with political, fiscal and administrative dimensions. The response will consist of a system for governance of forest resources covering territorial division; administrative responsibilities and coordination mechanisms at national, sub-national and local levels; identification of financial and physical assets and resources to be transferred; and, finally, sustainable management of forest resources.

Objectives and activities

In both countries the objective of technical assistance is to assist in designing the institutional framework (administrative, fiscal and technical arrangements) for the decentralized governance of forestry resources and in starting to implement decentralized governance. The project involves a phased implementation of decentralization, as follows.

Phase I: Situation analysis and first proposal for the decentralized governance of forest resources

Phase 1 comprises the following:

1 the administrative system for decentralized governance of forest resources:
- identification of key partners;
- inventory of decentralized governance experiences;
- analysis of the administrative system and of the implications for the decentralized governance of forests and natural resources;
- identification of communes' assets and patrimony;
- description and definition of administrative units as per current legislation; and
- description of authorities and mechanisms for accessing power and relationship with forests and natural resources management;

2 identification of technical responsibilities under the central government:
- identification of means and resources to be transferred and their mechanisms;

- identification of technical responsibilities of the state at the regional, provincial and commune level; and
- development of policy guidance for the different administrative units;

3 identification of implications in decentralized governance of forest resources:
- administrative decentralization;
- decentralization of forest management (technical aspects);
- decentralization of technical support;
- financial and budgetary decentralization; and
- identification of powers, responsibilities and stewardship;

4 setting objectives and goals for the decentralization of forest and natural resources:
- national level;
- regional level; and
- municipal level;

5 practicing decentralization:
- implementation of decentralized management of forests in a selected number of municipalities;

6 forest and natural resources management:
- adaptation of experiences into guidelines for a village management plan;
- preparation of guidelines for municipality forest and natural resources management plan;
- preparation of guidelines for regional management of forests and natural resources; and
- network for coordination, monitoring and stewardship of forests and natural resources;

7 proposal of the institutional arrangement for decentralized governance:
- institutional structure (central and local authorities);
- definition of functions, hierarchy, powers and responsibilities;
- preparation of post descriptions and organizational manuals and rules;
- preparation of planning and decision-making procedures; and
- definition of coordination mechanisms and networks;

8 preparation of cartography as a support for planning, monitoring and evaluation;

9 preparation of a system of evaluating the effects of decentralized management of forests and natural resources.

Phase II: Validating the proposal for decentralized governance of forests and natural resources

Phase II comprises the following:

1 feasibility analysis and comparison with current situation (costs and benefits);

2 procedures and criteria for monitoring and evaluation;
3 assessment of expected economic, social and environmental effects;
4 identification of and recommendation concerning required legal and institutional aspects;
5 interim plan for launching and starting up the implementation of decentralized governance of forests and natural resources in selected communes and regions.

Phase III: National programme for the decentralization of forests and national resource governance

Phase III comprises the following:

1 administrative, technical and legal guidelines for the decentralized governance of forests and natural resources;
2 plan for legal and institutional reforms at regional, municipal and village level;
3 plan for strengthening human resources;
4 budget and calendar for implenting decentralization at national level;
5 methodology for monitoring and evaluation.

Strategies

Since the first discussion at the beginning of the project, it has been clear that implementation must be carried out progressively in order to avoid a power vacuum. It has also been evident that in-country experience gained through participatory exercises and other activities can be an important starting point for implementing decentralization laws. At the lowest level, the decentralization scheme, following the principle of subsidiarity, will be based on village organizations and local knowledge. The goal of decentralized governance will be to improve individual and community welfare and the sustainability of those gains through sustainable forest management.

Observations and recommendations

Inclusion of forestry specialists

The movement towards decentralization has achieved a general acceptance by governments, donors and international organizations. In many countries because political decentralization has already reached maturity, actions are now being concentrated on fiscal and administrative decentralization. This mean that forest resources will be involved more frequently and more directly (for example, in Burkina Faso, Ghana, Mali, Niger, Sudan and Togo). This situation raises opportunities and challenges for using the experiences and knowledge developed with participatory forestry. To make the most of this experience, decentralization needs to be complemented with specialists from the discipline of forestry, particularly those specializing in forestry institutions, forestry economy and participatory forestry management. This interaction requires improved methods to reveal more clearly the implications of, and the relationship between, forest resources and institutional reforms for decentralized management with the countries' own national development strategies. In short, partnerships between the socio-economic groups leading decentralization with those concerned more with the orientation of people's welfare are the challenges for the forestry specialists facing a strong decentralization movement.

Civil society participation

The election of municipal authorities and the establishment of a governing body are the means for securing representation for political parties. However, more attention should be given to creating mechanisms for the formal participation of civil society organizations in the decision-making process of local governments.

Recentralization

A strong attachment to centralized systems exists among forest administration staff who, in most cases, were trained to run public administration under a top-down approach. Because decentralization is strongly supported politically, open resistance is not frequently expressed. However, special efforts are needed to avoid ending up with schemes that maintain centralized operation merely under new names and structures. Situations in which, for stewardship reasons, the state can take back (even temporarily) responsibilities from the local level should be replaced by mechanisms to deal with emergencies within the limits of decentralization. It is of particular importance to decentralize forestry resources where the stewardship involves many ministries. In the case of Burkina Faso, under the laws of decentralization, at least three ministries are directly responsible for forest resource stewardship: the Ministry of Environment, the Ministry of Territorial Management and the Ministry of Finance.

Human resources

Human resources, as is well known, are a weak point for decentralization in developing countries. The two cases reported here have an extremely high need to reinforce and further build institutional capacity. It may be acknowledged that national and international civil society organizations (CSOs) have been working to strengthen this aspect. However, in implementing decentralization of forests, special attention must be given to institutional strengthening at village and municipal level.

The knowledge and experiences of community forestry represent a very special complement for implementing decentralization of forests. The role that it can play in the sustainable use of forest resources and rural livelihoods is widely recognized. Furthermore, in some countries, community forestry has already moved beyond the experimental pilot stage.

Cost of public goods

The fact that natural resources are also a national patrimony calls for special norms for managing resources and for paying for the functions carried out by each actor. Should villages be paid for the public goods that are derived from their forest management? On the other hand, the communes' and villages' interests in financial gain should not damage national interests and the overall wealth of forest resources. There is a need to preserve the national public interest without using it as an excuse for reconcentration.

Unequal resource endowment

The dissimilarity in wealth of resources requires compensatory measures. Some regions have the richest resources and others might have the most serious problems due to degraded or overused resources. To deal with this situation, mechanisms for compensation and assistance to disadvantaged municipalities – especially during the first phase of implementing decentralization – will be required.

Financing sustainable forestry

Decentralization reforms, especially fiscal, are an opportunity for local authorities to begin to directly manage part of the national budget. In theory, such measures could increase investments in the conservation of the forestry sector. However, because of the many demands and problems faced by local government, there are few chances that new funds will be assigned to the forestry sector as a consequence of fiscal decentralization. Even worse, in some cases, the broadened responsibilities borne by local authorities after decentralization might become an incentive for raising more funds; in most cases, the main sources of revenues might be forests. This situation could result in an increased cutting of forests.

Potential negative effects of decentralization

Decentralization exercises have also had negative effects, especially when implemented in the absence of adequate mechanisms for accountability and people's participation. Preliminary examinations of decentralization in the forestry sector have identified situations in which, lacking proper management, decentralization can result in negative consequences. For instance, with decentralization there exists the possibility that local elites may gain power and control over local resources. Forest income is sometimes used by local authorities for funding public works, such as roads and urban infrastructure, in order to secure political goals such as re-election.

The possibility that the number of government employees will increase in decentralized forest administrations cannot be disregarded because in some countries the number of decentralized units is many times higher than under the centralized scheme. For example, Mali has some 700 municipalities to which forest resources and responsibilities will be transferred. Another reason is that in some countries the government is the main employer, and posts in the public sector are used as a means of paying back political clientele. Decentralization may also create new situations of financial stress for local government when central governments are not prepared to relinquish part of their traditional revenue at the same time that local governments charged with new responsibilities seek to increase theirs. This results in an increase in taxation on forests, which, in turn, stimulates illegal exploitation of timber. The context for forestry decentralization needs to be carefully assessed during the planning phase of forestry decentralization, and the implementation must be phased accordingly with the evolution of those dimensions.

Community and participatory forestry

Community forestry and decentralization share many of the premises that encourage the transfer of power, resources and responsibilities to people and local government. The knowledge, experiences and lessons of community forestry can serve as an important technical and managerial starting point and complement to decentralization when it arrives in the forestry sector. From a legal and administrative point of view, experiences with community forestry systems such as collaborative management are useful models for designing regulatory frameworks that harmonize formal and customary legal systems. Community forestry experiences are also of great help in developing new regulatory frameworks and institutional structures and in harmonizing modern government approaches with traditional ones.

Currently, the concept of community forestry, understood as 'any situation that intimately involves local people in forestry activity' (Arnold, 2001), has been changed to the more comprehensive idea of participatory forestry – the processes and mechanisms that enable those people who have a direct stake in forest resources to become part of decision-making in all aspects of forest management, from managing resources to formulating and implementing institutional frameworks. The participatory forestry approach can potentially play an important role at the policy formulation level for the decentralization of the forestry sector. Adherents of participatory forestry and its body of principles, knowledge, experience and methods can also interact more closely with economic and social analysts to make premises and models that strengthen local people's decision-making power more effectively. Participatory research also needs to strengthen its methods for demonstrating in quantitative and tangible terms the implication of its practice for people's welfare.

Challenges to the discipline of forestry

Important potential gains could be drawn from a partnership between forestry development efforts and the wider movement of decentralization. However, the discipline of forestry will need to provide the theoretical and technical evidence of the relevance of the forest sector to political, administrative and fiscal decentralization in a way that is understandable to the economists and politicians leading the movement. Greater emphasis needs to be put on documenting results, progress, objectives, goals and targets as points of departure for analysing the processes of decentralization in the forest sector.

The systematic documentation and objective analysis of the effects of decentralization of forest management must be urgently addressed in order to make an early selection of the strategies and methods that could bring better results to people and sustainability. A joint FAO–World Bank initiative could be developed for analysing and identifying strategic issues at the regional level, and for formulating forestry decentralization programmes that are complementary to those already under way in fiscal, political and administrative areas.

Conclusion

The objectives and assumptions of the benefits of decentralization are clear in economic, political and administrative areas. But in the forest sector, experiences are still too new to draw final conclusions on the effects of the processes of decentralization achieved to date. Many positive and negative effects have been identified, and it is clear that countries need to strengthen their analytical capacities in this area. At the same time, they need technical assistance for decentralization in the same way that they received assistance for political, administrative and economic reforms.

The objectives of decentralization are normally measurable and are defined in time for key economic and social aspects, such as economic growth, employment creation and productivity. The same cannot be said for the forest sector, where the objectives of decentralization are less specific or are even occasionally unknown. It is necessary to reach the same level of analysis and capacity for design in the forest sector.

References

Arnold, J. E. M. (2001) *25 years of Community Forestry*. FAO, Rome

Ayres, W. S. (undated) *Online Sourcebook on Decentralization and Local Development*. World Bank, Washington, DC, www.ciesin.org/decentralization/, accessed 11 October 2003

Camdessus, M. (1999) *Rationale for Decentralisation: The Online Source Book in Decentralisation*. International Monetary Fund Conference on Second Generation Reforms, www.imf.org/external/np/speeches/1999/110899.vhtm

IDB (Inter-American Development Bank) (1997) *Latin America after a Decade of Reforms*. IDB, Washington, DC

Litvack, J., Ahmad, J. and Bird, R. (1998) *Rethinking Decentralization in Developing Countries*. World Bank, Washington, DC

Mai, C. (1999) *Decentralized Rural Development and the Role of Self Help Organizations*. Workshop, Thailand, 4–6 November 1998, RAP Publication

Manor, J. (1997) 'The promise and limitations of decentralization, part 4'. Paper presented at Technical Consultation on Decentralization, Rome, 16–18 December

Rondinelli D., Nellis J., and Shabbir Cheema, G. (1984) *Decentralization in Developing Countries: A Review of Recent Experience*. Working paper no 581, World Bank, Washington, DC

Ter-Minassian, T. (1997) *Decentralising Government, Finance and Development*. International Monetary Fund, Washington, DC, September

UNDP (United Nations Development Programme) (2004) *Human Development Report – Millennium Development Goals: A Compact among Nations to End Human Poverty*. Oxford University Press, New York

World Bank (1984) *Reviewing the Evidence on Decentralisation*. World Bank, Washington, DC

World Bank (1998) *Fiscal Decentralisation*. World Bank, Washington, DC

World Bank (2000a) *Democratic Decentralisation*. Operation Evaluation Department, World Bank, Washington, DC

World Bank (2000b) 'Decentralization: Rethinking government'. In World Bank, *World Bank Development Report 1999/2000*. Oxford University Press, New York, pp 107–155

Chapter 6

Implications for Biodiversity Conservation of Decentralized Forest Resources Management

Jeffrey Sayer, Christopher Elliott, Edmond Barrow, Steve Gretzinger, Stewart Maginnis, Thomas McShane and Gill Shepherd

INTRODUCTION

Conservation organizations generally consider experience with decentralized and devolved forest management over recent decades to have been very positive. In some instances, natural forests have spread as a consequence of the institution of local management arrangements. But in many cases, decentralization has occurred too rapidly and has been subject to multiple and conflicting pressures. Some attempts have failed. One cannot assume that local communities will be able to re-establish traditional systems of forest management overnight after years of central government interference. Baseline data on environmental values are often lacking as a basis for evaluating the performance of decentralized forest management systems, particularly for biodiversity values. Devolved management systems have a vital role to play in achieving patchworks of different forest types in multi-functional landscapes. This chapter finds that, in certain circumstances, decentralization is a positive force for conservation; in others, it poses dangers. The issue is often not whether to decentralize, but how. The chapter offers general guidelines based upon our experience in Eastern Africa, Guatemala, Nicaragua and the Philippines.

DECENTRALIZED FOREST RESOURCES MANAGEMENT

The management and conservation of forests have often been a source of tension between powerful, centralized state authorities or the ruling elite and less powerful local communities. The history of forest governance and forestry as a profession reflects the underlying tension between the 'centre' and the 'local'. In theory, centralized forest institutions were established as the guarantors of the public goods and intergenerational values of forests on the assumption that local forest users would seek immediate and private benefits. The reality was often a struggle between different sectors of society for land and valuable timber and wildlife resources.

The way in which rights and assets have traditionally been partitioned reflects existing power structures. A model for forestry that evolved in Central Europe during the 18th century subsequently formed the basis of much forest law in the Asian and African tropics. This gave the ownership of land, timber and high-value game animals to the aristocracy or the state, and rights to lower-value products, such as dead timber, mushrooms and pasture, to the peasantry. The conservation movement that emerged when people began to attach value to dramatic landscapes, rare animals and plants and wilderness during the 19th century sought to add these values to those over which the state sought to exercise control (Schama, 1995; Rackham, 2001).

The modern conservation movement began with the establishment of national parks and similar protected areas, where outstanding natural features were placed under permanent state control. The removal of important natural areas from local control and their gazettement as state-controlled protected areas was the prevailing paradigm. The number and extent of such protected areas were often the principal criteria for the success of conservation programmes. Until recently, inclusion in the United Nations list of protected areas required that an area be placed under the jurisdiction of the 'highest competent authority' of the state. Protected areas were expected to have clearly defined boundaries and management plans. Until the relatively recent past, most governmental and non-governmental conservation organizations focused their efforts on the establishment of new national parks and equivalent reserves, or on supporting the improved management of those that existed.

Mainstream environmental groups have commonly advocated setting aside 10 per cent of all forests as inviolate protected areas. More radical groups would like to see all remaining near-natural forests given total protection. In 2003, the World Parks Congress in Durban, South Africa, celebrated the fact that the 10 per cent target had been exceeded. More than 12 per cent of terrestrial ecosystems worldwide are now included in some sort of formal protected area.

During the final decades of the 20th century, international conventions and processes were developed to give global stewardship to some natural areas. The World Heritage Convention, adopted in 1972, is a significant manifestation of the tendency towards global governance. Many people still advocate the designation of the world's high conservation forests as World Heritage sites under a form of global protection (Sayer et al, 2000).

Set against this tendency to centralize control and ownership of forests has been a tradition, deeply rooted in history, for the development of local rules and institutions to conserve and ensure equitable access to forests. Arrangements for conserving forests and other natural resources were among the forces that drove the emergence of local governments and democratic processes. In many parts of the world, traditional arrangements for managing forests have functioned successfully for centuries (Watkins, 1998). Modern foresters and conservationists often ignored these local forest stewardship arrangements and, in many cases, took actions that weakened or disempowered local forest managers.

Forest conservation by and for people

Formal conservation organizations rarely worked through traditional local forest management structures. Colonial conservation authorities in tropical countries largely ignored local people's rights. However, good managers have always recognized both the need to foster good relations with local people and the value of habitats outside the formally designated protected areas for many species of wild fauna and flora. Some conservation programmes allowed local traditional uses of certain resources within protected areas and sought to ensure that some benefits from the areas flowed to local people. There have been many attempts to create buffer zones around strictly protected core zones; sport hunting reserves around African national parks are a notable example. However, the prevailing paradigm has remained that of using state-controlled protected areas as the primary means of conserving wild nature.

The dominant role of central authorities in managing conservation began to be questioned during the 1970s and 1980s. First it became apparent that it was difficult to ensure the survival of protected areas in the face of opposition and resentment on the part of local land users. Second, it began to be recognized that much biodiversity depended upon habitats that could not realistically be included in totally protected areas. Third, much valuable biodiversity existed in areas of extreme human poverty, and conservation programmes that were indifferent to, or even exacerbated, this poverty were morally indefensible. These realities have been major issues of debate within the conservation community during recent decades. The theme of the World Parks Congress in Bali in 1981 was 'Parks and People'. The following decades saw a proliferation of initiatives to reconcile conservation with local development, many of them with limited success (McShane and Wells, 2004).

Since that time we have witnessed major efforts to involve local people in managing conservation programmes. We have sought 'win–win' solutions to conservation and development conflicts. This shift in emphasis among conservationists has come at a time when development practitioners have realized the benefits of empowering local resources managers and giving them rights to, and responsibility for, the resources upon which they depend. The force that drove the convergence between the conservation and development communities came from the realization that natural resources, including forests, were often the prime assets upon which the livelihoods of poor rural people depend. Official development assistance emerged as the major source of funding for forest conservation in the developing world.

The result has been a period of two decades when almost all conservation organizations – governmental, intergovernmental and non-governmental – have been struggling with a variety of approaches to decentralized management of all aspects of biodiversity. The results of this experience are generally considered very favourable for local livelihoods (see, for example, Poffenberger, 1989; Wiley, 2001; Barrow et al, 2002; Ribot, 2002, 2003). It is more difficult to assess the impacts of such schemes directly on biodiversity. There are unanswered questions about the relative importance of short-term poverty alleviation versus longer-term environmental sustainability (McShane and Wells, 2004). The relative values of biodiversity as a locally valuable resource and biodiversity as a global heritage constitute a difficult issue. People's views of the value of biodiversity and of the costs that they are prepared to incur for its conservation will depend upon their circumstances, culture and religious beliefs (Sheil, 2001). Nevertheless, abundant but mostly anecdotal evidence suggests that well-conceived and executed decentralization schemes have resulted in more extensive forests of mixed indigenous species in many localities where native forests were under severe threat. Devolved management has undoubtedly provided real biodiversity benefits.

The degree to which biodiversity conservation can be devolved or decentralized is very much a question of societal choice; there is no single correct formula. The balance of this chapter attempts to document experiences of the World Wide Fund for Nature (WWF) and the World Conservation Union (IUCN) in working with types and degrees of decentralization of forest biodiversity management over the past two decades. These decentralization processes rarely had conservation as an objective. We were often working in situations where forces outside our control were operating to change the nature of access and tenure rights. We were almost never operating in situations of orderly, planned and democratic transfer of rights and responsibilities from centralized institutions to decentralized ones. In this chapter we use the term *decentralization* to cover a broad range of transfers of the locus of decision-making from central governments to regional, municipal or local governments. We distinguish this from *deconcentration,* the process by which the agents of central government control are relocated and geographically dispersed. Devolution means the transfer of rights and assets from the centre to local governments or communities. All of these processes occur within the context of national laws that set the limits within which any decentralized or devolved forest management occurs.

Direct impacts of decentralization

The shift of management authority from the centre to the local level has sometimes had harmful effects on protected areas. These negative impacts have often occurred when the decentralization process was a result of a crisis and was therefore hurried, ill planned or entirely anarchic. Recent history of the protected areas of the remoter parts of Indonesia shows how dangerous it can be to decentralize control of globally important protected areas when no local institutions or funds are available to fill the management vacuum. Important protected areas in Indonesia have suffered illegal logging and encroachment during the recent period of transition of power over natural resources from Jakarta to the districts. The decentralization process in Bolivia is generally thought to have been positive for commercial forest management; but the decentralized authorities have shown less interest in protecting the broader environmental values of the protected areas. In Peru, responsibility for protected areas was retained by the central government, while forest management was decentralized. The decentralized authorities have challenged the logic of this; but for sites of national or global biodiversity value, the Peruvian solution may be optimal.

Local involvement in protected area management

Almost all protected area managers, and particularly international projects to support protected areas, now attempt to involve local populations in management. There are numerous attempts to ensure that benefits from the protected areas flow to local people. However, few cases qualify as true devolution of rights or assets. People are consulted, given options for employment and, in some cases, receive a share of entrance fees; but they rarely have any real control over the resources themselves. Critics contend that protected area managers are lagging behind forest managers in exploiting the full potential for giving real authority over resources to local people. The literature on integrated conservation and development projects contains many accounts of failed attempts to encourage local people to adopt environmentally friendly land uses in areas adjacent to protected areas (McShane and Wells, 2004). Much of the criticism of such projects focuses on the failure to genuinely empower local people or to give them usable rights or assets. Although much has been written about various approaches to sharing the benefits of protected areas, there are few examples where local people receive benefits that exceed the opportunity costs that they incurred when the protected area was created. The economic benefits of tourism still mainly accrue to tour operators and airlines. Little progress has yet been made in capturing the theoretical value of biological resources for the benefit of local people.

Protected areas

Protected forest areas, and the local management arrangements for them, take several forms.

Private nature reserves

Nature reserves operated by private foundations and non-profit conservation groups are common in industrialized countries. The costs of many of these privately operated reserves are met from entrance fees and the fund-raising activities of the sponsoring organizations. High-profile reserves with conspicuous rare species can generate profits that are often reinvested in other areas with less public appeal. Many such reserves generate local employment and enhance local incomes.

The few examples of similar private reserves in developing countries are concentrated in a small number of countries where special local circumstances make them profitable. Costa Rica, Ecuador and South Africa account for the vast majority of private reserves. The reserves in Costa Rica and Ecuador are successful because of the demand for rainforest and bird-watching tourism, mainly from the US. The South African examples mostly focus on a limited niche for high-cost safari tourism. The local private nature reserves that exist elsewhere are not a product of decentralization but of local enterprise.

IUCN categories V and VI areas

IUCN categories V and VI refer to conservation of protected terrestrial landscapes and marine areas (V) and sustainable use of natural ecosystems (VI), respectively; such managed resource areas are, in principle, compatible with decentralization. Some extractive use of natural resources is allowed, with restrictions that ensure the maintenance of environmental values. National forests in the US, where strict regulations ensure the maintenance of populations of important wildlife species, are a good example. The regional natural parks in France and the national parks in the UK are also examples. These categories of protected areas have a mixed record for biodiversity conservation.

The US national forests are considered by many people to have superior natural and wilderness values to many national parks, mainly because they have fewer visitors and less intrusive tourist infrastructures. Proponents of pristine wilderness criticize them because the extractive management sometimes conflicts with the conservation of certain rare species. In the European examples rights and assets are decentralized. However, these models have weaknesses when local economic interests conflict with the conservation of species of national or global value. Biodiversity conservation objectives might have been better met by the establishment of IUCN category I to IV protected areas. However, local interests would not have tolerated the loss of access to resources that category I to IV designation would imply. Categories I to IV are:

I integral protection (integral natural reserves/zones for wild nature);
II conservation of an ecosystem (national parks);

III conservation of natural elements (natural monuments); and
IV conservation for active management (habitat/species management areas).

The Brazilian Extractive Reserve concept has been widely praised as a potentially successful way of reconciling national or global nature conservation objectives with those of local people (Ruiz-Perez et al, 2005). A high level of management responsibility for resources is devolved to the local level, and natural values are conserved as a collateral benefit. Many conservation organizations have advocated the wide replication of the extractive reserve model, particularly in the tropical moist forest biome. Considering how enthusiastically the first extractive reserves were greeted by conservationists, it is surprising how little replication of the concept has occurred – little outside Brazil and almost none outside South America. Part of the reason lies in the special circumstances of the Brazilian situation: most of the value of the forest was in products that could be harvested in a non-destructive way, primarily latex and some other non-wood forest products, and timber was mainly high-value species that occurred at low densities. In other tropical regions the main marketable values of the forest lie in lower-grade timbers whose harvesting requires considerable disturbance. In most cases industrial-scale timber extraction requires investments in logging infrastructure and equipment that would be beyond the means of local people.

Nevertheless, many conservationists see category V and VI protected area arrangements as having a great deal to offer in developing countries. Some expect that major expansions in protected area systems will come from these categories. In most cases protected areas in these categories will be subject to more decentralized management than traditional national parks and other sites in IUCN categories I to IV.

Sacred forests and traditional areas

In many of the more densely populated parts of Africa and Asia there are large numbers of small forest areas that are protected by local people. The protection is often linked to religious belief systems, but the areas are also frequently of utilitarian or amenity value. Many of these areas have been protected for a long time and rigorous local rules determine any use that takes place. These sacred groves or forests are often the only areas of near-natural forest in densely settled agricultural landscapes. In at least some areas protection of these sites is more effective than that of small nature reserves established by government agencies in similar situations. In addition, many of these areas have been shown to support biodiversity values of considerable significance. In some parts of West Africa and India, small populations of rare mammals and birds have survived in these isolated areas. Formal conservation agencies often gave little attention to sacred forests; but recently, as all other forest areas came under greater threat, these areas have begun to attract more attention. In India, biodiversity inventories have been conducted, and measures are being taken to support local efforts to conserve the forests. Similar measures are needed in other countries. Sacred forests are examples of decentralized forest conservation activities that have proved resilient in the face of increasing human pressures.

BIODIVERSITY OUTSIDE PROTECTED AREAS

A significant shift in approaches to biodiversity conservation resulted from the realization that the vast majority of species are entirely or partially dependent upon non-protected areas for their survival (Halliday and Gilmour, 1995). The importance of conserving biodiversity in the managed landscape has now been widely recognized by national conservation agencies, intergovernmental processes (such as the Convention on Biological Diversity, or CBD, and the United Nations Forum on Forests, or UNFF) and especially by conservation non-governmental organizations (NGOs). Large-scale conservation planning and eco-regional or landscape-based programmes have now been adopted by all the major international conservation organizations. Forestland outside strictly protected areas will often need special management attention in order to achieve biodiversity conservation objectives, and many of these non-protected areas are coming under decentralized management regimes. What are the implications of this for biodiversity conservation?

Conservation organizations have thrown themselves enthusiastically behind devolved and decentralized management of forests outside protected areas. This has largely been motivated by the expectation that many forests in developing countries are more likely to be well managed and even to persist in the face of pressure for land clearing if they benefit local people. Conservationists believe that local management will tend to be for more diverse products and to be, in general, more environmentally friendly than the conventional timber orientation of central forest agencies. The reaction of conservationists is partly motivated by their scepticism at the prospects for the sustainability of large-scale industrial forestry.

Demand for land and forest products often makes total protection of forests impossible. In these situations, conservationists often view the various forms of local forest management as the least bad option. The forests that result may not be as rich in biodiversity as a natural forest; but at least forest cover is retained. Studies have shown that locally managed forests or agroforests often support a remarkable variety of plant and animal species (Michon and de Foresta, 1995; Fay and Michon, 2003).

Many authors have argued that local forest management is good for forest conservation (for example, Poffenberger, 1989; Fay and Michon, 2003). However, there is surprisingly little empirical evidence of what happens to biodiversity as a result of the process of decentralization. The following four case studies documenting the biodiversity outcomes of decentralization processes in Eastern Africa, Guatemala, Nicaragua and the Philippines have been selected from numerous examples in tropical regions. In most cases it is too early to tell whether the approach will be sustainable and yield the desired biodiversity outcomes. Methods for assessing and monitoring biodiversity are still inadequate to permit a scientifically valid conclusion; but the case studies suggest that in some situations, at least, decentralization processes have yielded positive biodiversity outcomes.

Eastern Africa

Most of the remaining closed forests in East Africa are small remnants in landscapes dominated by agriculture or fire-impacted grasslands. These forests have high levels of species diversity and many locally endemic plants and animals. Many have long been the focus of attention for conservation organizations, which in the past usually focused on the establishment of forest reserves, nature reserves or national parks under central government control. In the forest reserves there were often attempts to apply sustained-yield forestry under forest department regulations and supervision. During recent years the forests have suffered a great deal of degradation from uncontrolled logging, often sponsored by political elites, and from agricultural encroachment.

The forests are important sources of products for local people. A range of local management systems existed in the past that regulated the uses to which the forests were subject. Central government forest programmes marginalized many of these traditional management systems, especially during the colonial period.

Recently, there has been a resurgence of interest in traditional management systems and most of the countries in the region have decentralized control of management of forests to a greater or lesser degree. Most conservation organizations have aligned themselves with this decentralization tendency, and almost all now base their programmes in the region on local management paradigms. Consensus appears to be that under local management, forest areas are being maintained or extended, and it is proving easier to regulate the activities of non-local exploiters of the forests. Most observers attribute any difficulties to imperfections in the way that decentralization was handled, rather than a problem with the basic concept. Excellent overviews of decentralization in the region are given in Wiley and Mbaya (2001) and Barrow et al (2002).

One well-documented example relates to the Shinyanga region in Tanzania. A long tradition of the conservation of *ngitilis,* areas of dry forest protected by the communities as sources of fodder, fuel and other forest products, had existed. When many villages were relocated under the *Ujaama villagization* scheme from 1975, the *ngitilis* were largely abandoned, and the government promoted tree planting, mainly of exotic species. Beginning in the mid 1980s, government schemes began to support greater local decision-making in forest management and encouraged people, once again, to manage natural regeneration. Local memory of the *ngitili* system was still strong and there was demand for *ngitili* products. In 1998, Tanzania revised its forest policy and placed greater emphasis on participatory management and decentralization, which gave further support to the *ngitilis*. Village authorities are now empowered to enact by-laws, apply traditional rules and appoint guards to protect their *ngitilis*.

Between 1980 and 2001, 18,039 individuals were involved in managing *ngitilis*, with a total area of 78,122 hectares (ha) in 172 of the 833 villages in the region. The total extent of *ngitilis* increased from 600ha in 1986 to 250,000ha in 2001. The average size of the *ngitilis* is 2.2ha, and they range from 0.1ha to 215ha. About half of the *ngitilis* are owned by communities and half by individual households. The *ngitilis* are certainly making valuable contributions to local livelihoods. The fact that they are largely mixed forests of indigenous species suggests that they are a good option for biodiversity.

Many of the biodiverse forests of Eastern Africa are located on mountains. Climates and soils in these areas are good for agriculture, and population densities are intensifying pressures on the forests. Conservation programmes for these forests, from Ethiopia to Zimbabwe, have all been exploiting the advantages of decentralized and local management. Problems have occurred when central forest departments have resisted divesting real rights and ownership of these forests to local people. But, in general, conservation programmes based on local management are considered the best option for maintaining these forests and, by extension, their biodiversity (Wiley and Mbaya, 2001; Barrow et al, 2002).

Guatemala

In 1989 the Guatemalan government disbanded an authoritarian regional agency responsible for the natural resources of the remote Peten region and established the 3.1 million hectare Maya Biosphere Reserve. Within two years, it became apparent that the government's new 'tight-handed approach' was incapable of stemming the flow of landless peasants into the reserve, and that the reserve's significant conservation values were severely threatened.

As a response to what could be called an anarchic situation, a consortium of conservation and development NGOs, donors and progressive government officials created the Comite Consultivo Forestal to develop a decentralized community forestry concession model. This model's goal was to develop an economic constituency for maintaining forest and reducing forest conversion via natural forest management. The committee developed rules and regulations delineating rights and responsibilities to forest resources for communities, first, and private companies, second.

The government withstood substantial pressure by the traditional wood products industries and established the first community concession of 7000ha in 1994. Long-term management rights have since been granted for more than 400,000ha of natural forest to both communities and industries. Much of this forest is now certified by the Forest Stewardship Council. Both economic and conservation benefits have been substantial: forest conversion, fires and illegal logging have been dramatically reduced in the parts of the reserve under concessions (whereas they have increased in the centrally managed national parks, such as Laguna del Tigre and Sierra Lacandon), and employment and earnings in remote forest communities in the reserve have benefited from sustainable forestry activities.

In this form of decentralization, the government develops rules and regulations (with substantial local public input), and the communities then shoulder the responsibility of implementing them. Illegal logging, forest conversion and invasions are the responsibility of decentralized local managers. In a sense, the government has passed the brunt of the protection responsibility to the communities who obtain economic benefits from forestry. The model works.

Nicaragua

The north-eastern corner of Nicaragua is home to the Miskito and Mayagna indigenous groups, who have enjoyed an autonomous, democratically elected government since 1990. This independent governance system is based on traditional systems. However, until recently, the national forest service made virtually all decisions related to forest management and natural resource use.

In 1992, the government decentralized approval decisions for environmental impact assessments of all projects in the area. This model has been relatively successful. National funds were used to establish a regional environmental authority that has assumed responsibility for local environmental management. This success led to a new forestry law, approved in August 2003, which specifically addresses decentralization of federal responsibilities to local authorities. It provides for approval, control and supervision of harvest operations; public consultation and approval processes; channelling of revenues to local government; and policy development.

In response to this increasing responsibility, the indigenous government's Natural Resource Commission has established a Comite Consultivo Forestal comprising professionals working on forestry issues in the region. The committee is finalizing a comprehensive forest strategy to delineate priority activities, funding sources and partners to ensure that not only the national government but also donors and NGOs engage in activities that benefit the Miskito coast.

Such work is already bearing fruit: ten indigenous communities have initiated natural forest management on more than 75,000ha of lowland tropical and pine savannah forests. Community forestry businesses are slowly being established, commercial linkages are developing, Forest Stewardship Council certification is being pursued, and a generally favourable response by most members of the wood products industry is the result.

The Philippines

Sibuyan Island is one of the most important centres of biodiversity and endemism in the Philippines. At the heart of the island is the Mount Guiting-Guiting Natural Park. It is the only remaining mountain in the country that retains relatively intact habitats along its entire elevation gradient. The park covers some 16,000ha of strict protected area and an additional 10,000ha of buffer zone.

This near-natural area is home to some 50,000 people, more than half of whom live well below the government-defined poverty level. The population includes approximately 1700 indigenous people known as the Sibuyan Mangyan Tagabukid, who live in the upland areas of the island and rely upon agriculture, timber, non-timber forest products and hunting. The creation of the protected area deprived these people of access to timber and non-timber forest products. In response to the needs of the local people, a new decentralized management regime has been developed for the approximately 5000ha of ancestral land that overlaps with the protected area.

NGO advocates of indigenous people's rights, such as Anthropology Watch, the Legal Assistance Centre for Indigenous Filipinos and the Philippine Association for Intercultural Development, helped the indigenous people obtain community title to their land and develop sustainable management systems for natural resources within their ancestral domain. The hypothesis was that secure land tenure combined with provisions of capital and training in resources management could ensure sustainable use of biodiversity. Management responsibility within the ancestral domain of the Sibuyan Mangyan Tagabukid was decentralized from Philippine government authorities to the indigenous community themselves.

By securing tenure over a large tract of land, subtle shifts in power over resource use have become discernible. With secure tenure, the Sibuyan people can enforce independent decisions over the resources within their domain. When local governments, for instance, need timber to build schools, they are now obliged to negotiate first with tribal authorities regarding the amount of timber that can be extracted from inside the ancestral domain.

In the Philippines, tenurial claims by indigenous peoples have largely been ignored since the Spanish declared all non-cultivated land as public domain, belonging to the state. With no recognized legal documentation showing proof of ownership (save for oral history), the indigenous people were squatting on lands that they had been occupying since before the creation of the Philippine state. The promulgation of the Indigenous Peoples Rights Act in 1997 was considered a landmark victory in the struggle to correct centuries of injustice. The process of determining tenure rights involves community consultations; ethnographic research; censuses; boundary identification and delineation; mapping; public notifications; a review; and, finally, the approval of community title by the Board of the National Commission on Indigenous Peoples.

In January 2001, four years after starting the tenure process, community title to the 7905ha ancestral domain was approved. A natural resources management plan has been developed and adopted by the indigenous community. The plan describes the Sibuyan priorities and commitments to sustainable resources management and identifies the resources needed to protect and develop their ancestral domain. In a review conducted in 2001, respondents observed reductions in illegal activities (logging and fish poisoning) and increases in reforestation and the planting of root crops. The opening of new agricultural land was reduced. The central authorities lacked the resources and knowledge to enforce environmental laws; but with the aid of the communication equipment provided by the project and increased awareness of forest laws, the Sibuyan have conducted patrols and apprehended timber poachers. Decentralized governance and secure land tenure have thus resulted in more effective management and law enforcement by the local people.

Restoration of degraded lands

Large-scale central government programmes to restore degraded lands have often been criticized by conservation organizations for having unfavourable social and environmental impacts (Sayer and Changjin Sun, 2003). Such schemes tend to promote vast monocultures of a single, often exotic, species of tree and are often implemented with little regard to the values that local people might attach to the land. Restoration programmes that are more responsive to local conditions frequently involve the decentralization of responsibility to local communities. In China, people have responded to the devolution of responsibility for degraded sites by establishing plantations of bamboo and trees (Ruiz-Perez et al, 2003). The plantations are primarily aimed at meeting market demand for fibre and fruits and may also be more favourable for biodiversity than the degraded lands they replace. There is evidence from some sites in China that degraded vegetation retains a considerable diversity of indigenous plants and animals, which these plantations displace. But other studies suggest that plantations can be surprisingly rich in biodiversity and that simple modifications of management practices can further enhance this biodiversity value (Buck et al, 2003). The International Tropical Timber Organization (ITTO) has recently published guidelines for forest restoration and rehabilitation that favour biodiversity and other environmental values (ITTO, 2003).

Armed conflict and civil strife

Recent studies have shown the extent to which armed conflict and civil strife are prevalent in forest areas in the developing world (Price, 2003). Such conflicts can undermine or destroy the institutions charged with forest management. There is some evidence that decentralized management arrangements for forests may be more resilient in the face of such disturbance than centralized management regimes. In Zimbabwe the community-managed wildlife areas are said to have survived the current period of civil unrest better than the national parks. Local people clearly saw a greater interest in protecting resources that they could exploit for their own benefit than in protecting resources that were under central management.

A contrary situation has been noted in Liberia and the Democratic Republic of Congo, where well-established national parks, some of them World Heritage sites, appear to have survived the turmoil of the past decade better than one might have expected. The sites had a high international profile, and the international community channelled resources to conserve them even during periods of conflict. Rebel factions in both countries have engaged with international conservation groups to ensure the protection of these forest protected areas – perhaps to establish their credentials as potential participants in legitimate and responsible governance bodies after the conflict.

FEDERAL AND CENTRALIZED GOVERNMENTS

In federal systems, conflicts have arisen over sites of biodiversity importance between provincial governments that tended to be responsive to resource extraction industries and central authorities that had longer-term conservation objectives. Australia is a notable example where the state governments have been much more responsive to issues of rural employment and primary industry interests, whereas the federal authorities, responding to the urban electorate in Sydney and Melbourne, have favoured conservation. This debate has been especially intense on the issue of World Heritage listing for some protected areas. Listing brought the sites under more direct federal control and in some instances was strongly resisted by the state governments. Similar issues have arisen in Malaysia and Canada.

NATIVE LAND CLAIMS

During recent years, conservationists have often found common cause with indigenous peoples' groups, whose traditional systems of land management have been recognized as favouring biodiversity. However, there are also instances where indigenous peoples are making claims upon land that is already gazetted in category I to IV IUCN protected areas. The native land claims often find constitutional support, since the laws and treaties upon which they are based predate those that led to the creation of protected areas. This issue may become more significant in many developing countries as local people react to legal systems and protected areas established by colonial or centralized dictatorial regimes. Claims for exemption from wildlife and fisheries regulations, especially concerning hunting seasons, bag limits and size restrictions, have often brought native peoples into conflict with government conservation agencies. These conflicts are likely to be exacerbated under decentralized regimes.

RECOMMENDATIONS

Conservation organizations generally consider decentralized and devolved forest management over recent decades to have been very positive. But the record shows failures, too: decentralization occurred too rapidly; central authorities retained control of the most valuable assets; hasty or opportunistic decentralization left institutional vacuums and lack of accountability. One cannot assume that local communities will be able to re-establish traditional systems of forest management overnight after years of central government interference. In those cases where devolved management did not really work, we now understand why. If it had been done right, better outcomes would have been achieved.

Much of the experience of decentralization has occurred in situations where all governance institutions, both central and local, are weak. Decentralization or devolution alone will not solve these problems. Most programmes to devolve forest management to communities and local authorities have been driven by the need to improve local livelihoods. Conservation organizations have often become involved after the process was already engaged. Often, few baseline data on environmental values are available for evaluating the performance of decentralized forest management systems.

The basis for the optimism of conservation organizations is the contrast with centrally managed forest areas. In many less developed countries, central institutions have had great difficulty in conserving forest environments, and forests are disappearing rapidly. After the institution of local management arrangements, however, natural forests have stabilized and even spread. Although central control of the extensive protected areas that are the cornerstone of forest conservation needs to be retained, devolved management systems have a vital role to play in achieving the complementary patchwork of different forest types in multifunctional landscapes that are essential to address broad conservation goals.

The general conclusions of this chapter are that, in certain circumstances, decentralization is a positive force for conservation; in other situations, it poses dangers. The crucial issue is often not whether to decentralize, but how. Based on our experience, the following guidelines suggest the conditions under which decentralization can favour biodiversity conservation:

- Devolved and decentralized management schemes have often led to the maintenance or extension of areas of species-rich indigenous forests and have thus provided biodiversity benefits. In many situations, local management has proved more successful in achieving this than centralized management.

- Decentralization must be planned and negotiated; crisis decentralization risks leaving an institutional vacuum.
- When the benefits of protected areas and other conservation measures accrue mainly at a broad public or global level, it may be unwise and unrealistic to assume that decentralized management will be effective. Category I to IV IUCN protected areas should, in general, be retained under central government control.
- Local managers must have legitimacy and be representative of local resource users; a large degree of local democracy may be a prerequisite for any sort of decentralization to work (Ribot, 2002, 2003).
- Assets, rights and power must be transferred to local managers, not just responsibility.
- Resources to support conservation programmes must be available to decentralized institutions.
- Economic benefits must accrue to decentralized managers and these must be equal to or exceed the opportunity costs forgone in order to ensure biodiversity conservation.
- When the value of conservation activities accrues mainly to external constituents, decentralized managers must be given some regulatory or financial incentive.
- Direct environmental payments are often necessary to offset local costs and opportunity costs of programmes to conserve biodiversity whose value accrues to the 'global community' and not directly to local people.

Note

The authors are grateful to Madhu Sarin, Leonardo Lacerda, Simon Rietbergen and anonymous reviewers for helpful comments on earlier drafts of this chapter.

References

Barrow, E., Clarke, J., Grundy, I., Jones, K.-R. and Tessema, Y. (2002) *Analysis of Stakeholder Power and Responsibilities in Community Involvement in Forest Management in Eastern and Southern Africa*. IUCN, Eastern Africa Regional Office, Nairobi, Kenya

Buck, A., Parotta, J. and Wolfrum, G. (2003) *Science and Technology: Building the Future of the World's Forests and Planted Forests and Biodiversity*. IUFRO occasional paper no 15, IUFRO, Vienna

Fay, C. and Michon, G. (2003) 'Redressing forest hegemony: When a forest regulatory framework is best replaced by an agrarian one'. Paper presented at CIFOR Conference on Rural Livelihoods, Forests and Biodiversity, Bonn, May

Halliday, D. and Gilmour, D. A. (1995) *Conserving Biodiversity Outside Protected Areas: The Role of Traditional Ecosystems*. IUCN, Gland, Switzerland, and Cambridge, UK

ITTO (International Tropical Timber Organization) (2003) *Guidelines for the*

Restoration of Degraded Forests. ITTO, Yokohama
McShane, T. and Wells, M. P. (2004) *Getting Biodiversity Projects to Work: Towards More Effective Conservation and Development*. Columbia University Press, New York
Michon, G. and de Foresta, H. (1995) 'The Indonesian agro-forest model', in Halladay, P. and Gilmour, D. A. (eds) *Conserving Biodiversity Outside Protected Areas: The Role of Traditional Ecosystems*. IUCN, Gland, Switzerland, and Cambridge, UK
Poffenberger, M. (ed) (1989) *Keepers of the Forest: Land Management Alternatives in Southeast Asia*. Kumarian Press, West Hartford, CT
Price, S. (2003) *War and Tropical Forests: Conservation in Areas of Armed Conflict*. Food Products Press, New York
Rackham, O. (2001) *Trees and Woodlands in the British Landscape*. Phoenix Press, London
Ribot, J. (2002) *Democratic Decentralization of Natural Resources: Institutionalizing Popular Participation*. World Resources Institute, Washington, DC
Ribot, J. (2003) 'Democratic decentralization of natural resources: Institutional choice and discretionary power transfers in sub-Saharan Africa'. *Public Administration and Development*, vol 23, no 1, pp53–65
Ruiz-Perez, M., Belcher, B., Maoyi F. and Xiaosheng Y. (2003) 'Forestry, poverty and rural development: Perspectives from the bamboo subsector', in Hyde, F., Belcher, B. and Jintao X.(eds) *China's Forests: Global Lessons from Market Reforms*. Resources for the Future, Washington, DC
Ruiz-Perez, M., Almeida, E., Dewi, S., Lozano-Costa, E. M., Ciavatta-Pantoja, M., Puntodewo, A., de Arruda Postigo, A. and Goulart de Andrada, A. (2005) 'Conservation and development in Amazonian extractive reserves: The case of Alto Jurua'. *Ambio*, in press
Sayer, J. A., Ishwaran, N., Thorsell, J. and Sigatty, T. (2000). 'Tropical forest biodiversity and the World Heritage Convention'. *Ambio*, vol 29, no 6, pp302–309
Sayer, J. A. and Changjin S. (2003) 'Impacts of policy reforms on forest environments and biodiversity', in Hyde, F., Belcher, B. and Jintao X. (eds) *China's Forests: Global lessons from Market Reforms*. Resources for the Future, Washington, DC
Schama, S. (1995) *Landscape and Memory*. Harper Collins, London
Sheil, D. (2001) 'Conservation and biodiversity monitoring in the tropics: Realities, priorities and distractions'. *Conservation Biology*, vol 15, no 4, pp1179–1182
Watkins, C. (1998) *European Woods and Forests: Studies in Cultural History*. CABI, Wallingford, UK
Wiley, L. A. and Mbaya, T. (2001) *Land, Peoples and Forests in Eastern and Southern Africa at the Beginning of the 21st Century: The Impact of Land Relations on the Role of Communities in Forest Management*. IUCN Regional Office for Eastern Africa, Nairobi, Kenya

Part II

Country Cases

Chapter 7

Decentralization of the Forestry Sector: Indonesia's Experience

Wandojo Siswanto and Wahjudi Wardojo

Introduction

The Indonesian Ministry of Forestry is committed to combating illegal logging, controlling forest fire, rehabilitating and conserving natural resources, restructuring the forestry sector and strengthening the decentralization process of forestry. Those priorities are designed to guide forestry stakeholders in solving problems at the national scale. The National Forest Programme, the ministry's approach to decentralization, requires multi-stakeholder dialogue at all levels. Land tenure, however, remains a problem: some local authorities sell community land to generate revenue, and landless farmers engage in illegal practices in utilizing resources. Thus, landownership does not always mean that a community can manage its holdings sustainably.

Reform in Indonesia

Indonesia is facing an ongoing process of fundamental change and democratic reform, including developing a stronger governmental relationship with civil society. This transitional phase, together with the economic and social impact of the 1997–1998 economic crises, has created a fragile and volatile situation. The reform movement has encouraged improvements in political life, such as efforts to maintain national sovereignty, increase the role of all stakeholders and reduce government dominance. However, such reforms are a long-term process because they require not only legal acts, but also changes in the institutional set-up and management of public affairs.

The reform movement that started in 1998 has pushed the implementation of governmental decentralization in a relatively short time. Since independence in 1945, the governance system has been changed no fewer than six times, back and forth from centralized to decentralized. The current system, adopted as a political commitment, is decentralization, based on the issuance of Act No 22 of 1999, which went into effect on 1 January 2001. However, the legal formulation was carried out rapidly and under transitional circumstances following the fall of the previous regime; hence, the act is incomplete and imprecise.

A few months after the issuance of Act No 22 in 1999, Forestry Act No 41 was passed. The implementation of these two acts has created problems for the parties responsible for managing the forest and forest resources. Problems are particularly severe between the central government and local government agencies because of different interpretations of the term *decentralization.* Lack of understanding, coordination and synergy between central, provincial and local governments has led to misperceptions in identifying responsible actors in forest management. Therefore, distribution of authority and roles has become an area of conflict among the parties concerned. In addition, problems have emerged because of a lack of transparency and accountability in managing and maintaining forest resources as a life-support system, rather than only as a source of timber.

Misperceptions regarding the implementation of decentralization in the forestry sector by local authorities were mainly caused by forest managers' short-term orientation, which is related to the terms of office of local government administrators: five years. As a result, the forest is perceived as a source of timber that generates local revenues needed for local development. Timber-cutting licences are issued locally without sufficient regard for sustainable principles, leading to rapid deforestation.

Decentralization of the forestry sector should, however, be viewed as a positive effort to bring public services closer to the people and manage forest resources in a sustainable manner for the community's welfare. Therefore, differences should be discussed in constructive dialogues in a transparent manner to yield solutions agreed by stakeholders.

The minister of home affairs has declared that the decentralization process should be conducted in a four-stage process: initiation, installation, consolidation and stabilization. The four steps have become a baseline for all development sectors to implement decentralization within their respective authority.

In order to handle the current complex problems of the forestry sector, the Indonesian Ministry of Forestry has developed a plan: Five Priority Policies on Forestry. The five commitments of the ministry are

1 combating illegal logging;
2 controlling land and forest fire;
3 rehabilitation and conservation;
4 restructuring the forestry sector; and
5 strengthening the decentralization process of forestry.

The five priorities are designed to guide forestry stakeholders seeking solutions to problems at the national scale. An important tool for such an approach to decentralization is the National Forest Programme, a policy framework that requires multi-stakeholder dialogue at all levels. In addition, two cross-cutting programmes are being conducted: social forestry and human resources development.

The objective of decentralization in the forestry sector is to bring public services closer to the Indonesian people by providing local communities better access to forest resources in their areas. Access, however, must be in line with the principles of sustainable forest management, which must be planned and widely disseminated to local people. In this way, in the future, well-managed forests can support better living conditions and improve the welfare of the Indonesian people.

IMPLEMENTATION OF DECENTRALIZATION

Act No 22 of 1999 (local government) and Act No 25 of 1999 (financial balance between central and local governments) require decentralization in most aspects of governance. A transformation from centralized to decentralized governance is not an easy move given that *decentralization* may be interpreted differently by different parties. In the forestry sector, some local governments have interpreted decentralization as total freedom to do whatever they want with the forest resources in their region. Such a misinterpretation obviously endangers the very existence of forest resources. Nevertheless, as a political commitment, decentralization has reached a point of no return and must proceed. Smoothing and accelerating the decentralization process in the forestry sector has now become an urgent priority requiring intensified communication, coordination and collaboration among the parties concerned.

Decentralization means the handing over of political, financial and administrative authority from central to local (district or city) governments so that the government can facilitate and guarantee better public services for the people. The success of decentralization depends upon the availability of proper institutions and qualified human resources in implementing all aspects of decentralization, as well as other supporting factors, such as infrastructure, technology, information access, personnel and institutional capacity in planning, and – more importantly – the proper distribution of natural resource benefits.

Recognizing the spirit of autonomy, implementation of forestry decentralization should not be trapped in the polemic of disputes over authority. However, it should be oriented towards achieving sustainable forest management consistent with the vision of national forestry development, on the one hand, and promoting people's welfare, on the other. These two dimensions of forestry development could be achieved through establishment of proper coordination and understanding among the parties involved, at both central and local levels. Administrative boundaries, therefore, should not be a barrier in forest management; in fact, forests and forestry should become a kind of glue, since forests can significantly benefit regional, national and international development in terms of economic, ecological and socio-cultural factors.

Acts and regulations

As a result of different perceptions and understandings of decentralization, conflict over natural resources management has remained, particularly with regard to regulations. Policy gaps between Act No 22 and Act No 41 need to be resolved to clarify responsible parties and agencies in achieving sustainable forest management at central and field levels. However, since the two acts are already in effect, a synergistic approach to the two, within the framework of national sovereignty, becomes the issue.

Basically, Act No 22 authorizes district governments to execute all government tasks except those related to national and international issues, such as foreign affairs, national defence and security, justice, fiscal and monetary matters. However, according to Act No 41, utilization of natural resources, including forests, natural resources conservation and standardization are still the responsibility of the central government. The statement that 'management of natural resources situated in the region is under authorization of respective governments and they become responsible in maintaining the sustainability of the environment according to existing acts and regulations' leads to different interpretation and disagreement on natural resources management, including forests.

As part of the decentralization process, the Ministry of Forestry issued Decree No 05.1 of 2000 to decentralize the issuance of small-scale forest concession licences that permit the utilization of timber coming from land clearing. Implementation of the decree led to an uncontrolled number of licences that tended to encourage forest degradation. Hence, in June 2002, the government issued Regulation No 34 of 2002, withdrawing Decree No 05.1 of 2000. Since then, disagreement between district and central governments has emerged and created considerable debate among forestry stakeholders.

Regulation No 34 of 2002 (forest management and forest management plan formulation) and No 35 of 2002 (reforestation funds) were issued as implementing regulations of Forestry Act No 41. Both refer to other relevant acts, such as Act No 22, as well as Act No 18 of 2001 and Act No 21 of 2001 (special autonomy for the provinces of Aceh and Papua, respectively). According to Regulation No 34, all development activities undertaken by all government agencies, including local government, must promote the spirit of 'good governance', meaning that local government should take not only the authority but also the responsibility for conducting development activities in a transparent and accountable manner. Since forestry development has many implications for the economy, ecology and culture, forest degradation and forest act enforcement are duties shared by all levels of government.

As part of the forest management system, authority to issue forest utilization licences has become a focus in decentralization disputes. The mechanism for granting forest utilization licences is clearly defined under Regulation No 34, including a coordination mechanism among central, provincial and district governments. However, misunderstandings emerged regarding whether forest product utilization within natural or plantation forests should be approved by the minister of forestry. The issue triggered disagreement and the mechanism is perceived as recentralizing authority. In many areas, the issuance of permits for timber utilization and small-scale forest concessions by local authorities is still occurring. The situation leads to unsustainable practices of forest harvesting and causes severe damage to forests, the environment and ecosystems. Uncontrolled utilization of forest products has occurred both upstream and downstream in watershed areas and has even exceeded district, provincial and national boundaries.

Although the authority to issue licences for forest product utilization is now with the minister of forestry, technical recommendations from the head of the district are required. It is expected that through this mechanism, proper monitoring can be developed and maintained. The line of responsibility among stakeholders is also clear if, for instance, negative impacts occur during execution of this licence. Hence, in order to achieve better coordination, a clear delineation of roles and responsibility between central, provincial and district or city governments, as well as with state-owned companies, is crucial. Additionally, Regulation No 34 governs the utilization of forest and forest concessions that have been granted to the company and state-owned forest company.

Tenure conflict

In this era of decentralization, land tenure has become a crucial and disputed issue. Some *Adat* ('customary') communities contend that forests should be owned and managed based on the history of forest occupation. In addressing this issue, the government sets some policies regarding forest and land governance based on Act No 41. The policy on forestland is aimed at avoiding further degradation and fragmentation. Under this policy, the communities are becoming involved in forestry development and granted the rights to manage the forests for their livelihood provided they do so sustainably, without sacrificing the environment. In many cases, because of economic pressures, local communities transfer landownership to other parties to gain quick cash. The landless farmers then begin to engage in illegal practices when utilizing resources. Thus, landownership does not always ensure that the community can increase its long-term income.

Financial arrangements

District revenue as indicated in Act No 22 comprises four major components:

1. revenue called *Pendapatan Asli Daerah*, which is generated within a local jurisdiction;
2. balance of fund or central government transfers;
3. loans; and
4. other sources.

Balance of fund or central government transfers consist of local shares of land and building tax, fees for land and building acquisition and natural resource revenues; general allocation funds (*Dana Alokasi Umum*); and specific allocation funds (*Dana Alokasi Khusus*). Specific allocation funds, provided to cover expenditures for specific needs, include the reforestation funds (*Dana Reboisasi*). Other sources include grants and emergency funds. Act No 25 (balancing of funds between the central and regional governments) stipulates that the general allocation funds allocated to regions (that is, provinces, districts and municipalities) should be no less than 25 per cent of the domestic revenues stipulated in the national budget (*Anggaran Pendapatan dan Belanja Negara*). Of this, 10 per cent goes to the provinces and 90 per cent to the districts and municipalities.

Of the revenues collected from natural resources – namely, from the forestry, general mining and fishery sectors – 80 per cent goes to the respective regions (provinces, districts and municipalities) and 20 per cent to the central government. Compared with other natural resource sectors, the regions' share of forest-derived revenues is significant. Only 15 per cent of revenue from oil goes to producer regions; 85 per cent goes to the central government. Similarly, only 30 per cent of revenue from gas is returned to the producer regions; 70 per cent goes to the central government.

The reforestation fee is collected from concession holders as non-tax revenues and can be spent only for reforestation and rehabilitation of degraded forests. As stated in Regulation No 35, utilization of the reforestation fee is allocated for the following activities under loan schemes, except for conservation and protection purposes:

- replanting degraded production and protection forests and conservation areas, except nature reserves and the core zone of a national park;
- reforestation, particularly for community or private forest, forest conservation-related activities, and watershed conservation for agricultural systems;
- enrichment planting within production and protection forests; and
- supporting activities, including conservation area management, prevention and suppression of forest fire; management of forest boundaries; monitoring and surveillance; fining; seed development; research and development; education and training; and extension as well as empowering local communities in forest rehabilitation activities.

The collected reforestation fee is divided into two portions: 40 per cent goes to the timber-producing regions and 60 per cent to the central government, with the expectation that this 60 per cent will be focused on forest rehabilitation and forestry development in non-timber-producing regions, including forestry development support activities. It is suggested that a proper mechanism and distribution of revenue from the natural resources sector needs to be developed in a more transparent manner.

FIVE PRIORITY POLICIES

Considering the current complex problems in the forestry sector, the Ministry of Forestry has developed and launched its Five Priority Policies on Forestry for immediate action, to be used as a reference and guideline for forest management and forestry development in the provinces and districts. One policy concerns the need to strengthen the decentralization process in the forestry sector. Decentralization of the forestry sector is prioritized in order to build common perceptions and understandings among central, provincial and local stakeholders. Even though this is not an easy task, Indonesia believes that progress can be achieved through concrete efforts and good strategy.

Constraints

The decentralization policy in the forestry sector was aimed at handing over roles and authority for forest resources management from the central to local governments. However, its implementation was slow, and some conflict among parties arose because of the complexity of problems and misunderstanding of the forest management model in the field.

The problems and constraints in implementing decentralization in the forestry sector are demonstrated by obvious uncontrolled forest degradation during the era of decentralization. The annual rate of forest degradation recorded increased from 1.6 million hectares to more than 2.1 million hectares between 2000 and 2003. Moreover, other causes of deforestation include unsustainable timber exploitation and conversion of forestland for agriculture, estate crops and other purposes. The following problems are considered the main handicaps in implementing decentralization in the forestry sector:

- Different perceptions and low levels of understanding of decentralization among most forest management authorities persist. Local governments were unprepared in terms of institutional and human resource development, as well as their capacity for managing forests.
- The hierarchy between central and local governments, both province and district, was interrupted after autonomy was launched. Therefore, it is necessary to redesign Act No 22 of 1999 (local government) for better implementation in terms of transparency and harmony.
- Some stakeholders propose revising Act No 22 because of its exaggerated implementation of decentralization. Members of the People's Consultative Assembly (MPR) issued Decision No IV/MPR/2000 concerning policy recommendations on implementation of decentralization, and the House of Representatives issued a letter in February 2001 proposing amendment and improvement of the act.
- Controversy remains about local government regulations at the district level on small-scale concessions and licences for timber utilization. The

current policy actually contradicts Regulation No 34. The short-term orientation of district governments to increase their local revenue from timber-harvesting activities is clearly seen.

- Claims by local communities, particularly traditional *Adat* communities, to own forestland have grown since autonomy and decentralization began.
- Conflicts in managing forests in the field have been routinely encountered because of incomplete instructions for the implementation of Act No 41.

Although various programmes and projects have facilitated a forum for dialogue and discussion in terms of the National Forest Programme process, constraints remain because of weak institutional capacities, little knowledge and understanding of decentralization, as well as lack of coordination among stakeholders.

Strategy

Considering all constraints in implementing decentralization in the forestry sector, the following efforts and strategies are required:

- Implement the National Development Programme and Annual Development Plan according to existing acts and regulations. This should be followed by proper implementation of the five priorities of the Ministry of Forestry at all levels.
- Promote dialogue and effective communication among forestry stakeholders, including government agencies at all levels, legislative institutions, universities, non-governmental organizations (NGOs), the private sector and representatives of *Adat* communities, in implementing the National Forest Programme.
- Draft relevant government regulations as an elaboration of Act No 41 and its implementing regulations. Three of ten planned government regulations have been issued by the president; the remaining seven are under discussion.
- Develop standards and criteria for sustainable forest management, and at the same time increase the capacity of institutions as well as human resources in the forestry sector.
- Build a positive image in forestry development by establishing two-way communication, particularly with the mass media.
- Promote cooperation with donors (international institutions) for sustainable forest management.
- Enforce the forestry act and other relevant acts as they pertain to illegal forestry practices.

An important recommendation of the World Forestry Congress 2003 in Quebec, Canada, is that forestry should be put on the political agenda of each respective country. That agenda can then be channelled in the wider context of poverty reduction, economic and national development, as well as cultural integrity and respect, global climate and decentralization – issues on which forests and forestry have significant impacts. Inter-sectoral, governmental and multi-stakeholder dialogues need to be established and maintained properly. The tool designed for such an approach is the National Forest Programme, which is progressing slowly.

National Forest Programme

The concept of the National Forest Programme has been introduced, and the policy is in the initial stage of formulation. With decentralization, multi-stakeholder participation is recognized as important in achieving sustainable development; thus, there is a heightened awareness of the necessity of involving stakeholders in the policy process.

The National Forest Programme was accepted as the reference framework for sustainable forest management. Conservation and development of all types of forests are expected to be effective in tackling existing environmental problems. The programme is also recognized as a tool in achieving sustainable forest management, in particular, and sustainable development, in general. The challenge lies in how this process can be implemented effectively. A wide range of interests among stakeholders and a knowledge gap among them are factors that need special attention in the participatory process.

In order to ensure transparency in the process and equity in benefit-sharing, an appropriate mechanism should be established. Furthermore, a conflict resolution scheme is required, given the diverse backgrounds and interests of stakeholders in forest management. The National Forest Programme is expected to be an effective tool to build common perceptions among stakeholders, taking into consideration the rising awareness about decentralization and the lack of understanding of forest resources management. The programme could be a way of creating a synergistic spirit among all forestry stakeholders as they work together cooperatively towards sustainable forest management.

Yet, the concept of the National Forest Programme as a process for establishing a comprehensive cross-sector framework is relatively new to most Indonesian forestry stakeholders. There is a prevailing perception that such a programme will be national and centralistic in nature. Clarification of the meaning of the programme's concepts and principles has been followed by establishing an institution to undertake the programme's development, with immediate action plans. This was initiated in 1999 with the involvement of donor countries through the Consultative Group on Indonesian Forestry.

In 2001, a working group on the National Forest Programme identified a list of strategic issues pertaining to Indonesian forestry development. These strategic issues were presented in several multi-stakeholder consultations across the country to solicit responses and suggestions, as well as to trigger local initiatives. The results will be the basis for developing the major themes for Indonesian forestry, including a national forest statement.

The concept of the National Forest Programme has been shared through consultation and workshops in the provinces. During 2001, workshops identifying forestry strategic issues were conducted in ten provinces of Indonesia. The wide range of interests of different stakeholders and the knowledge gap encountered among stakeholders were carefully considered in a participatory manner. Several issues of common understanding emerged from the workshops, ranging from technical to conceptual and from local to national. Stakeholders shared a common perception on some fundamental issues, such as the concept of sustainable yield and ecosystem-based forest management, the desirability of a review to harmonize forestry legislation, and the need to reduce logging and rehabilitate degraded forests and lands. The need for a benefit-sharing mechanism among stakeholders, including upstream–downstream compensation mechanisms, was also discussed, as well as the important role of local communities in promoting their own welfare.

Discussions on macro planning within the context of the National Forest Programme have also involved various stakeholders. The discussions have led to recommendations, such as using the programme as a tool for developing comprehensive policy-making and a mainstream policy framework to accommodate the complex problems of forestry development. Developing a partnership mechanism with international institutions and NGOs to support forest and land rehabilitation and a conservation programme was also recommended.

In line with Indonesia's decentralization policy on forestry, it is believed that the National Forest Programme could become a useful instrument to build common understanding and develop coordination among stakeholders, and ultimately to come up with an agreement that wins the commitment of all stakeholders. Since its adoption in 1997, the programme has been used for the promulgation of decentralization policies under which all services and management of forests were to be transferred from the central to the local level.

Given the importance of consultation and discussion in the provinces and districts, open-mindedness and transparency are particularly necessary to implement decentralization successfully.

Conclusion

The implementation of decentralization in the forestry sector is in its initial stages, and progress is slow because of constraints imposed by conflicting policies and regulations. Different perceptions and understandings of the spirit of decentralization can be resolved gradually by establishing proper dialogue mechanisms, as well as by building consensus and capacity. The National Forest Programme could be used as an effective framework. Preparation of policy and regulations, as well as guidelines, criteria and indicators, needs to be accelerated in order to keep decentralization of the forestry sector on the right track.

Indonesia will continue to organize consultations with local stakeholders and promote the importance of forests to present and future generations. Recent natural catastrophes, which were due, in part, to improper forestry development and management, have reinforced the message, and it is believed that local stakeholders will gradually implement forest management in a more sustainable manner.

Chapter 8

Forests and Decentralization in Switzerland: A Sampling

Christian Küchli and Jürgen Blaser

Introduction

Switzerland has a long experience with decentralization and federalism. During the mid 19th century, large-scale clear-cutting of Alpine forests started in connection with high demand for timber from colonial countries and industrialization in Switzerland. The disastrous floods occurring simultaneously were linked to this clear-cutting. In response, the newly constituted Swiss Confederation passed a forestry law in 1876. Brief case studies of four sites in the Bernese Oberland illustrate the development of the interplay among government stakeholders at the national, cantonal and community levels, and between public and private forest owners. Stable relationships between public and private forest owners and the cantonal forest service emerge as a precondition for long-term forest management; but at the same time, changing conditions of the socio-economic context constantly make it essential to seek a new equilibrium among the three levels of government, as well as among the public and private forest owners and civil society.

Decentralized forest management and conservation in Switzerland

Switzerland comprises 26 cantons and about 3000 political communes. The country grew out of treaties between valleys, townships and other geographical units and was finally organized in its current form in 1848, when the constitution of the Swiss Confederation was enacted. Forest resources management and conservation have since been shared by the national (federal) level in Bern, the regional (cantonal) level and the local (commune and community) level, the latter being the main forest owners in Switzerland. Several elements characterize decentralized forest management and conservation in Switzerland:

- the importance of capacities as well as decision-making power and accountability at the local level;
- strong guiding provisions at the federal and cantonal level and the subsidiarity principle;
- the development of a 'cooperative federalism', in which all three levels – local, cantonal and federal – have clear responsibilities;
- the length of time required to develop cooperative federalism;
- the need to build institutional capacity from the federal to the local levels, which historically was strengthened by financial contributions from the federal to the cantonal forest agencies;
- the principle of 'coupled contributions', by which the costs for providing environmental services such as protective functions are shared among all three levels;
- the positive relations between local political decision-making and centralized technical guidance and funding at cantonal and federal levels;
- the need for stable relationships between public and private forest owners and the cantonal forest service as a precondition to achieving long-term forest management goals;
- the constant need to seek a new equilibrium among the three levels of government, as well as among the public and private forest owners and civil society, especially under the current economic pressure of globalization and the related cross-sectoral influences.

One rationale for conducting the Interlaken workshop on Decentralization, Federal Systems of Forestry and National Forest Programmes in April 2004 was to facilitate an exchange of experience between countries that have long been decentralized, such as Switzerland, and newly decentralizing countries, such as Indonesia – the two hosts of the workshop. Switzerland's long experience with decentralization and federalism was presented to workshop participants through a background document (von Arb and Zimmerman, 2004) and four field excursions to illustrate the variation that characterizes the Swiss governance landscape. In Switzerland, as elsewhere, decentralization and centralization are two sides of the same coin. It is the way in which these two features of any governance system interact that ultimately determines how forests are being conserved, managed and used.

Historical overview

During the mid 19th century, the already-fragile ecological balance from centuries-long overuse and overgrazing in upstream forests collapsed when extensive flooding affected the Bernese Oberland and most other parts of Switzerland. Large-scale clear-cutting of Alpine forests during the early 19th century had accelerated the disasters.

One main reason for deforestation in the Bernese Oberland was a fundamental change in the governance system in the 1830s. For many centuries, the powerful city of Berne had held sovereign rights to forests in the Oberland, while the local communities only had use rights. However, prior to the creation of the confederation, the canton began to hand over forest property rights to the communes and, subsequently, also to farmers. This liberalizing process coincided with powerful market forces generated by the high timber demand of a growing economy in Switzerland and Europe – the main driving force for extensive clear-cuts. The colonial powers, particularly France and Holland, were important in generating demand. Hence, the gates to timber export were opened wide; but both legal provisions to protect the forests and a forest service of competent specialists were lacking. By the mid 1850s, in some parts of Switzerland forest cover was less than 10 per cent, compared with almost 30 per cent today.

The disastrous floods, which even affected cities in the river valleys and lowlands, triggered the federal government's involvement in the forest sector, eventually leading to the enactment of the first Federal Forestry Law of 1876. The new law, written in a highly political climate, had to accommodate the legitimate interests of the self-governing Alpine cantons. The solution was a federal legal framework that included three important overarching regulations:

1 Forests could no longer be reduced in size.
2 Cut areas had to be replanted within three years after harvesting.
3 Owners of public forests, such as the community-level governments, were required to develop forest management plans that specified proposed uses of the resource.

In order to assist forest owners and to help enforce these rigid provisions, the federal government provided compensation as an incentive for the reforestation of protective forests. The federal law also obligated the cantons to employ trained foresters. Thus, foresters' roles expanded not only to enforcing forest legislation, but also to acting as forest management consultants to local forest owners. This approach allowed foresters to take fully into account the cultural and biological diversity of different geographical areas.

The cantonal forest laws had to comply with the federal law where it specifically addresses the particular needs of a canton. The innermost legislative layer included locally established regulations to address the design of transparent community institutions for making management decisions, distributing benefits, resolving conflicts and defending community interests against the selfish desires of outside forces and local elites.

Local capacities and local-level decision-making power and accountability, combined with the strong guiding provisions at the federal and cantonal level, have proven critical in preventing deforestation and further forest degradation. Moreover, well into the 1970s, sustainable forest management and timber production provided a significant stimulus to local economies and also satisfied national and international demand for products and services.

Since the enactment of this first federal forest legislation, the repartition of rights, responsibilities and duties has had to be renegotiated through several revisions of the law. Federalism in Switzerland is an ongoing process of repeatedly finding a new equilibrium among the central state, its member cantons and public and private forest owners. Globalization is increasing the pace of this process. Moreover, the principles of the 1992 Rio Earth Summit also broadened the scope of forest policy in Switzerland towards better incorporating concerns for biodiversity and forest environmental services provided by forests. Addressing forest issues as a dimension of managing global public goods is a new concept in forest management that still needs to be properly introduced at decentralized levels.

Issues in a federal system: Four examples

The total forested area of Switzerland in 2000 was about 1.1 million hectares (ha), of which approximately 70 per cent is owned by communes and communities. The forested area of the canton of Bern is 176,000ha, of which 51 per cent is private forest, mostly owned by smallholders with less than 5ha of forest. This is far above the national average of 27 per cent private forests. Although the Bernese Oberland (the area around Interlaken) is dominated by communal forests, private ownership is characteristic of the foothills region of the Emmental and the central plateau (Mittelland) of the canton. This is a result of the reconstitution of forest ownership that took place during the 1840s (Küchli and Stuber, 2004). In addition to geographical and ecological differences, the diversity of landscapes also translates into socio-economic, cultural and political diversity.

Using four sites in the Bernese Oberland as examples, we discuss issues of decentralization in the forest sector. The four sites were visited during the Interlaken workshop to give participants a direct understanding of decentralized forest management in Switzerland. The cases emphasize interfaces among government stakeholders at national, cantonal and community levels, and between public and private forest owners. Each demonstrates the interplay of local and regional factors in relation to the Bernese Oberland and national issues previously discussed.

Site 1: Rugen-Mürren – from government to governance

Rugen and Mürren provide an historical overview of the development of forests and forestry at the local level, starting during the 19th century. A variety of closely related processes over the past 200 years, both within and outside the forestry sector, shaped the nature of forestry, including the development of energy sources as a major driving force.

Decentralization of forest ownership and capture of resource access by the elite

The decentralization process and reconstitution of forest ownership that began during the 1840s initially had disastrous impacts. The period prior to the passing of the Federal Forestry Law of 1876 was the most difficult in the history of forestry in the Alpine regions. In social terms, it was marked by a seizure of the resource by the rural elite and the exclusion of the economically and socially disadvantaged segments of the population. In ecological terms, it was characterized by the clear-cutting of large tracts of Alpine forests. Economically, it meant resource depletion in the short term. After a transitional period of some 20 years, however, the clarification of forest ownership rights provided an important foundation for sustainable forest management. The focus was on multifunctional timber production and environmental protection, a characteristic of the management of Swiss forests over the past 120 years.

Involvement of the federal government

Events such as the flood disasters during the middle of the 19th century stimulated the involvement of the federal government in the forestry sector, leading to the enactment of the 1876 law prohibiting reduction of forest area and requiring replanting and forest management plans. Even the owners of private forests could not sell wood that had not been marked by the forestry service.

The law made only modest provisions for support, however. Subsidies were provided as an incentive for reforestation of areas with protective functions. But just ten years after the introduction of the Federal Forestry Law, a parliamentary commission concluded that it had not led to the progress desired because federal subsidies were too low, particularly for afforestation. Moreover, Alpine cantons were not in a position to develop the required forestry expertise on their own. As a result, the introduction of sustainable forest management in communally owned forests and effective control of timber harvesting in private forests were long delayed.

In 1902 the federal law was revised to include more effective financing measures. The main aim was still to protect people and property from natural disasters. But it was now easier to pursue ecological and economic aims as well. The principle of combining protective and productive functions in the same area was born.

Respect for local autonomy as a basis for close-to-nature forest management

Swiss forest history is marked by latent local resistance to intervention by the canton, and cantonal resistance to intervention by the federal government. Yet, this resistance helped to maintain motivation and initiative at the local level. In general, local knowledge about sound management approaches in local forests was thus preserved. The remarkable diversity of Swiss forests and the approach of close-to-nature forest management are not least the result of this phenomenon, which could be called 'silvicultural federalism'. Decentralized forest management, however, is not a guarantee for close-to-nature silviculture. There are still many traces today of inadequate silvicultural provisions during the past, such as single-species plantations of Norway spruce (*Picea abies*) that were introduced some 130 years ago in mixed hardwood ecosystems.

Capacity-building, local people and the forest service

Jeremias Gotthelf, a popular writer of the Emmental Valley during the 19th century, once remarked that a forest warden knew approximately as much about silviculture as a hen knows about playing the violin. Karl Kasthofer, the first forester of the Interlaken region appointed in 1806, waged a long struggle on behalf of the education and training of foresters. Already in 1818, he maintained that 'what applies to the lowland forests of Germany cannot be applied to Swiss Alpine forests'. He therefore committed himself to the development of a Swiss forestry curriculum and academic training in forestry. In the view of Kasthofer, indigenous knowledge was to be a part of this curriculum. He urged: 'We should provide training to our own people ... but also listen to their opinions and take account of their observations.' In 1855 the Federal Institute of Technology in Zurich added professional training in forestry to its curriculum.

It took a long time for the people of Switzerland to cease regarding foresters only as police and to see them as professional advisers. Even though Swiss communes maintained their efforts to remain autonomous, foresters began to play an important role in many; the optimal management of community forests is not possible without professional advice. Foresters usually had to live and work for decades in the same district, where they often played a prominent role in local politics. They also had to understand people as well as trees, and they needed salaries that would at least support a middle-class lifestyle.

Incentives from the lowlands and socio-economic change

The fate of forests and forest management in the Bernese Oberland has always been influenced by demands originating in urban areas of the lowlands. For centuries, these demands were mainly for timber and fuelwood. Only after the disastrous floods during the 19th century did urban society in the lowlands commit itself to the preservation of Alpine forests. The financial incentives of the confederation, as provided by the Federal Forestry Law, enabled the cantons to operate more efficient forest services and to undertake torrent control and afforestation in catchment areas. During the 1960s, when the Alps became a recreation area for mass tourism, resources were again made available for a new generation of technical measures and afforestation to ensure better protection against avalanches and rock fall. Favourable changes in the socio-economic context played an important role in promoting and accelerating sustainable forest management. This is one of the central lessons learned from nearly 200 years of Alpine forest history in Switzerland.

Cooperative federalism: From government to governance

During the course of the 20th century, the federal government has continually increased incentives, especially for the preservation of forests, primarily through subsidies. This mechanism requires intensive cooperation among the confederation, cantons and communes. Together with financial incentives, the regulative instruments concerning forest preservation (for example, the ban on deforestation and clear-cutting) and the detailed prescriptions concerning cantonal and communal forestry organization have led to the development of considerable institutional capacity at all three levels. The top-down government approach of the 1876 Federal Forestry Law has gradually been refined into a governance approach, with close cooperation among government and civil society.

New phase of decentralization: Retransferring responsibility to lower levels

Several factors external to the field of forestry have influenced forest policy during recent years, most prominently the declining economic viability of forest enterprises because of low timber prices and the diminishing of public funds for forest development activities. These factors have led to a discussion of the distribution of forestry-related tasks between the state and the private sector, as well as among the three levels within the confederation. In Switzerland, this discussion is currently under way within the participatory process of the Swiss National Forest Programme, in which all important stakeholders are taking part, among them representatives of cantons and communes. The results so far suggest that political responsibility for forests will be partly retransferred from the federal level to the cantonal and commune levels. This does not exclude the confederation's role in retaining its commitments to certain partial tasks, like the maintenance of protected forests or the preservation of biodiversity.

Contemporary problems and solutions in mountain forests

The protective function against natural hazards (avalanches, rock fall, erosion, floods and landslides) is predominant in multifunctional forests and is critical for the survival of the population. Ever-increasing mobility has further heightened the need for the permanent provision of protective functions. Yet, forests as a living protective structure need constant care to guarantee the maintenance of all their functions. Steadily rising costs (particularly wages) and falling incomes from wood sales have led to a shortage in forest funds. Public money is needed to fill this deficit. There is a certain contradiction here: on the one hand, the confederation and cantons (the 'public') expect forest owners to perform more services; on the other hand, the financing and actual provision of these services are being left to the regional or local level. In terms of governance this means that the vertical line – confederation–canton–commune – is being weakened and must be replaced by cross-sectoral partnerships at regional or local level. This includes cooperation among forest owners, regional cooperation among communities, associations of communes, and regional policy and cross-sectoral cooperation.

Local forest owners must now make use of all possible measures for erosion and landslide control, for forest protection and for bark beetle control; they must focus on areas where public interests predominate, undertaking the absolute minimum and exploiting all possible ways of generating income to finance forest enterprises, so that they can continue to maintain ecosystem functions. In some areas, management needs to be discontinued (for example, tending and weeding of natural regeneration). Income-generating opportunities include outdoor events, adventure, and camps for apprentices and managers. Financing by sponsors is a new option that is being considered by forest owners.

Site 2: Brienz – cooperative federalism to fight natural hazards

The site of Brienz helps us to understand the interface between the national and the cantonal levels. In a mountain country, protective forests are of paramount public importance. The protective function is generally of regional, rather than only local, interest since mountain forests regulate water and sedimentation further downstream. Hence, all three levels – communal, cantonal and federal – are involved in managing mountain forests sustainably.

The Brienz Torrents (*Brienzer Wildbäche*) geographically belong to the eastern Bernese Oberland and are in the territory of the communes of Brienz, Schwanden and Hofstetten. *Brienz Torrents* is, in fact, a generic term for six torrents and their catchment basins on the steep southern slopes between the Brienz Lake (altitude 564m) and the crest of the Brienzer Rothorn Mountain at an altitude of 2350m.

The area is renowned for its natural catastrophes, some of them with disastrous results for the local population. The topography, geology and climate in the region create conditions for torrents, debris flow, over-bank sedimentation, avalanches, rock fall, rock avalanches, landslides and erosion. In some places, these processes coincide and influence each other. Avalanches can, for instance, transport logs and rock material into the torrent channels, thus blocking them in places and causing further problems.

Historical antecedents

During the Middle Ages, large areas of forest were cleared in the catchment basins of the Brienz Torrents to create alpine meadows and hay fields. The conifer belt (spruce) between 1500m and 1800m altitude was particularly affected. The timberline was pushed down from nearly 2000m to 1400m–1600m in altitude. The absence of forests had a detrimental effect on run-off and on the bed load regime. The consequences were soon felt – the torrents caused destruction in the valley.

The torrents present the most significant danger for the valley. In 1499 the village of Kienholz was destroyed. In 1797, simultaneous overflowing of three torrents destroyed 37 houses. The most recent catastrophe occurred in 1896, when enormous depositions (120m wide at the front and 2.5m to 4m high) blocked the main road, the railway line and large areas of cultivated land.

Forward-looking inhabitants of the village of Hofstetten recognized the correlation at an early date: in 1599 they planned a prohibition of forest utilization on the mountain. There was, however, strong opposition to this ban, and the clearing of forestland was again permitted by a 'high public court of arbitration'. Following disastrous floods during the first half of the 19th century, the federal council charged experts with an inquiry on mountain forests; the results of their investigations gave no pleasing picture of the situation. Consequently, the federal council urged the cantons to adapt their laws to stop forest destruction. When this advice was not sufficiently heeded, legal provisions were made at the national level. Article 24 of the Swiss Constitution of 1874 gave the confederation the superintendence over the hydraulic engineering and forest police in the high mountains.

Public interest in protection measures

The potential danger of the Brienz Torrents and the damage that they could inflict created the need for protection measures to reduce the high risks. The local population alone would not have been capable of carrying out and financing extensive hydraulic engineering works and forestry protection measures after the catastrophic flood of 1896. The wave of sympathy generated throughout Switzerland paved the way for a federal relief programme. A parliamentary motion to urge the inhabitants of the village of Schwanden to emigrate by giving them financial aid was rejected by the federal assembly. A clear majority favoured enabling the inhabitants of Schwanden to remain on their native soil, and the federal assembly granted maximum financial contributions for protective structures in the torrents.

Despite the generous level of federal aid, the communes still did not have the financial means to carry out major works. The canton of Bern then stepped in, buying the entire area destined for afforestation, and the cantonal council gave the cantonal government the right to expropriate community lands to simplify negotiations with landowners. The canton subsequently purchased large parts (690ha) of the upper catchment basins of the Brienz Torrents and became the director of works for the forestry projects. The superior public interests of the confederation and the canton involved protection of both inhabited areas (housing development policy) and transportation network (roads and railway lines).

Site 3: Boltigen – local organizations and the cantonal forest service

The commune of Boltigen illustrates the interface between the commune and the cantonal level and the functioning of the forest community of Adlemsried – one of several forest communities within Boltigen. The political and administrative processes within the community of Adlemsried still follow traditional forms of democracy.

Between 1870 and 1999, the population of Boltigen shrank from approximately 2000 people to 1500 because of emigration to urban areas and smaller families. Most people work in agriculture, with about 40 per cent also having another income source. Whereas in 1980 only 30 per cent of the people worked outside the commune, by 1990, 54 per cent did. The commune covers 7701ha, of which 2000ha (or 27 per cent) is forest, with an additional 5000ha in meadow and pasture. Sixty-seven per cent of the forest is community owned and 33 per cent is private.

The commune of Boltigen comprises eight traditional communities, which date back to the 14th century. During the 16th century, these developed from co-operatives into political administrative structures. Adlemsried, one such traditional community, has a history of self-sufficiency because of its geographic isolation. The community has two parts: the *Bäuert*, or territorial administrative body, historically responsible for political and forestry aspects, and the commons (*Allmend-Gemeinde*), which is responsible for managing common pastures. Both parts are built on a longstanding historical democratic foundation. Today, however, these communities have lost their political importance and focus now on forest and commons management, whereas other responsibilities, especially social ones, have been transferred to the modern political commune of Boltigen.

The community is composed of homeowners with a documented house right or, more literally, a 'fireplace right'. If a house is sold, the former owner loses rights and membership in the community. These rights traditionally include the following:

- the right to graze sheep and goats on the commons;
- the right to graze cows during the summer season on community pasture land (one third cow per house right; cow rights may be leased or transformed into the right to a piece of arable land);
- community forest rights (an annual lot of firewood, including fencing wood and timber for the repair of buildings); and
- the right to vote in the general assembly.

The community members also have duties:

- statutory (unpaid) labour for maintaining roads and paths and clearing after timber harvesting;
- compliance with the decisions of the general assembly; and
- a share in the administrative costs of the community.

The community's forest, which covers 214ha, is locally divided into forest for commercial timber use, protected non-commercial forest and protected forest on steep slopes. The latest management plan was written in 1982 and was provisionally extended, unchanged, in 1992. It allows for an annual allowable cut of 500 cubic metres.

One of the most dramatic events during recent years was the Lothar storm of December 1999, when one quarter of the timber stock was destroyed. Although timber prices have been very low, there has been a substantial effort to harvest these fallen trees to forestall an infestation of bark beetle in the remaining stands, particularly in the protective forest area. House-right holders' statutory labour was used in this effort, as well as contributions from a contractor, a civil protection service and the women from an ice hockey team. A total of 16ha of seriously damaged forest were converted to forest reserve status, which the community will not harvest for at least 50 years.

Forest management is under the control of an executive commission, two members of which are in charge of routine matters; but important decisions are taken by the general assembly. Forest workers are employed by the timber buyer or the community and are paid by the cubic metre. The forest ranger, who is partly on the payroll of the canton, has an important role in supervising all forestry matters, including timber marking and bark beetle eradication efforts, without charge to the community. Other activities of the forest ranger are paid for by the commune, such as the measurement of timber for sale, and planting and tending trees.

Site 4: Emmental farm forests – resources management in private forests

The Emmental is both a geographical entity and a political district in the eastern part of the canton of Bern. It borders with the canton of Lucerne and is drained by the River Emme and its tributaries. Private ownership predominates in Emmental: more than 90 per cent of the forests (18,068ha) are in private hands, compared with the national average of 27 per cent. Private individuals, however, on average own just 3ha of forest.

The Emmental forests delivered enormous amounts of fuelwood and charcoal to the steel mills in the industrial centres outside the valley at the beginning of the 19th century. Livestock grazing in the forests, common at that time, also played a role in degrading forests by destroying the regenerative capacities of the trees. Until late in the 19th century, large tracts of forestland were converted into meadows and intensive pasture, largely in the service of cheese production. An ever-increasing need for timber and fuelwood for farms and industry heavily degraded the remaining forests.

The devastating floods that occurred during the middle of the 19th century created the legal bases that brought a fundamental change in forest development, particularly in Emmental. Until 1920, there were enormous efforts in reforestation all over Switzerland. The greatest increase in forest cover was observed in the nearly entirely deforested Emmental: large afforestation projects were carried out in the mountainous Napf area and in the catchments of the Ilfis, a main tributary to the Emme.

This was made possible thanks to a small agricultural revolution. Simple technical improvements such as the construction of pits to collect animal urine significantly increased the supply of fertilizer. Milk and cheese production soared as a result. Natural re-growth of forest due to the abandonment of marginal agricultural land increased. In the middle Emmen, which includes the Napf area, forest area increased between 1860 and 1980 by approximately 5000ha, to 7350ha; in the commune of Trub, the increase during the past 90 years amounted to 40 per cent, up to 3250ha. About 20 per cent of this area increase is due to natural re-growth and not to forest plantations.

Silver fir (*Abies alba*) and beech (*Fagus sylvatica*) are the predominant tree species in the district's *Plenterwälder,* or single-stem selection forests – with trees of all ages and diameters, achieved through harvesting of single old and mature trees. About one third of the forest area is managed as selection forest, through guided natural regeneration. Another third of the forest area is in a stage of transition towards selection forest. The remaining third comprises even-aged stands.

Selection forests can be managed only with shade-tolerant species. This management method has numerous advantages for the forest owners, who can wait for good timber prices. Owners can also save costs because planting or stand tending is not necessary. Most of the time, forest management consists of some liberation thinning and the monitoring of growth of middle-aged trees, as well as creating favourable regeneration conditions, generally through single-tree harvesting of trees of all sizes. Such selection forestry best fulfils all major forest functions, including the productive and protective functions, as well as biodiversity conservation and recreational aspects.

In terms of timber wealth, the Emmental forests have an average standing volume of 520 cubic metres per hectare and an annual increment of 12.4 cubic metres per hectare. The annual harvest is, on average, 9.7 cubic metres per hectare. The Emmental already has among the highest forest growing stock in Europe, and this stock will rise over the coming years because of a decrease in harvesting. The forest service would prefer to increase harvesting to the sustained-yield level, or to reduce the growing stock to an adequate level in order to regenerate some of the old-growth forests. There is, however, insufficient processing and marketing capacity in place; more importantly, timber prices are low and barely cover production costs.

As opposed to the Bernese Oberland, where community forests predominate, in the hilly region of Emmental, wealthy farmers traditionally own large private forest plots. These private forests have been maintained by special hereditary customs. By tradition, the youngest son inherits the entire farmstead, thus preventing its subdivision. As a result, many forests in Emmental have developed into silvicultural gems barely distinguishable from well-managed community forests. The main species of the original natural forests – silver fir, Norway spruce and broadleaf species such as beech – have been maintained in the forests over the centuries and are still being managed selectively.

In lower parts of Switzerland, in contrast to the Emmental, many private forests were clear-cut for speculative purposes soon after forest property was decentralized during the 1840s. Forest plantations that developed primarily into pure spruce stands largely replaced the original broadleaf forests. On the basis of this experience, Swiss forest experts came to the conclusion that communal ownership was generally better suited for sustainable forest management than private ownership. This perspective is now widely accepted in the current context, as management of forest resources requires integral and larger-scale planning and activities over a long period. In addition, a certain minimal area is needed for rational forest management. According to an evaluation conducted nearly 50 years ago, the privatization that took place in the wake of the reconstitution of forest ownership of the 1840s was 'one of the worst things that could have happened to the forests in Switzerland' (Grossmann, 1949).

The issue of private ownership of forests is of central interest in Emmental. Private forests are often found on the best sites. These sites are characterized by a high percentage of fast-growing spruce, producing large trees that are prone to windfall. Damage from windfall can be observed all over Switzerland. The same storm that devastated Boltigen in January 1999 affected 10 per cent of Swiss forests and felled 12 million cubic metres of wood in a single day.

Farmers have traditionally been the most important owners of private forests. But with the concentration of agricultural land, stimulated by economic change since the 1960s, today an ever-smaller number of the 250,000 private forest owners in Switzerland are farmers. New forest owners are urban citizens, many not concerned with the management of their forest holdings.

The main responsibility of the forest service (divisional forest office for the Emmental), particularly of the forest rangers, is to guarantee sustainable management of forests through adequate application of silviculture and timber harvesting by the forest owners. The forest service advises the district's 6200 private forest owners and coordinates on-the-job training of young farmers – the future forest owners – at the agricultural school. In addition, the forest service gives expert advice to communes and non-forestry officials on issues relating to landscape management and forest legislation.

The Emmental forest service is currently developing regional management plans in cooperation with forest owners, communes, representatives of nature conservation groups, hunting and sporting clubs, and others. The objective is to identify the specific interests in forests so that the forest service can take necessary measures more purposefully in protective forests and can open up forest areas for recreation and nature conservation, within its limited means. The new element is the involvement of all kinds of stakeholders in a preliminary stage of landscape and forest resource planning. This constitutes a completely new approach, even in the highly decentralized and democratic country of Switzerland.

Conclusion

As the Swiss examples show, the challenge to manage forests sustainably is a permanent and ever-evolving task that does not start or stop with the degree of centralization or decentralization in decision-making and forestry funding. With changes in society, forest management also changes. Keeping the capacity to adapt forestry to new socio-economic contexts and economic situations is, and remains, the major challenge in the sector, even in such a highly decentralized country as Switzerland.

Note

The workshop field trips described in this chapter were conducted by Christian Küchli; Rudolf Zumstein, divisional forester of the Interlaken Region; Erica Zimmermann, Swiss Agency for the Environment, Forests and Landscape (BUWAL); Ueli Ryter, natural hazards specialist, Forest Service of the Canton of Bern; Beat Stucki, forestry consultant and executive committee member of the community of Adlemsried; Christian von Grünigen, divisional forester of the Frutigen-Obersimmental-Saanen region; Jürgen Blaser and Walter Marti, divisional forester of Emmental.

References

Grossmann, H. (1949) 'Forstgesetzgebung und Forstwirtschaft in der ersten Hälfte des 19. Jahrhunderts, 1803–1848, Beih. Zeitschr, Schweiz'. *Forstver*, vol 25, p63

Küchli, C. and Stuber, M. (2004) *Forest Development in the Swiss Alps: Exchanging Experience with Mountain Regions in the South* [CD-ROM]. Swiss Agency for Development and Cooperation SDC, Bern

von Arb, C. and Zimmermann, W. (2004), *Federalism: A Characteristic Element of Swiss Forest Policy*. Swiss Federal Institute of Technology, Zurich

Chapter 9

Decentralization of Forest Management in Bolivia: Who Benefits and Why?

Pablo Pacheco

Introduction

In Bolivia, decentralization and the involvement of municipal governments have had both direct and indirect effects on forest management and forest users' livelihoods. Although in the Bolivian model of decentralization of forest management municipal governments have more room than during the past to intervene in forest management, some of the major decisions regarding forest resources allocation and use are still made at the central level. Decentralization has reshaped the institutional mediations influencing natural resources use; but its outcomes are primarily influenced by social and economic conditions operating at the local level, such as local power relationships and property rights, which have been little affected by ongoing decentralization. Decentralization has contributed to improving social equity in access to forest resources, distribution of economic benefits from such resources, and reinforcing sustainable practices of forest management, although this does not happen in all cases. The challenges to progress in decentralization involve adjusting the design of the model of decentralization itself, as well as with overcoming other factors influencing it, such as local elite capture and formalization of property rights.

The move towards decentralization

During the early 1990s, countries around the world initiated a process of transferring some responsibilities for social services to local governments. The approaches taken in Latin America vary with the political economy of natural resources use in each country, the existing institutional systems linked to forest management, along with existing power relationships and government priorities (Larson, 2003). The outcomes of decentralization are also diverse, and they depend not only upon the type of responsibilities that municipal governments have been granted, but also upon a range of social and economic factors, such as local power relationships, transparency in policy decision-making and dominant property regimes. Although decentralization tends to challenge some of the institutional mediations influencing natural resources use (such as the sources of authority for forest management monitoring), it barely challenges other conditions influencing natural resources management (such as property regimes). Nevertheless, the higher level of democracy that decentralization promotes tends to have multiplier effects on natural resources management. This chapter supports this argument using the Bolivian case.

Compared with other countries in Latin America, Bolivia has adopted a more aggressive model of decentralizing forest resources management. This chapter seeks to explain the contextual factors influencing the path of Bolivian decentralization, depicts the new distribution of functions regarding forest resources, and determines its implications for local forest users.

LOWLAND FORESTS AND FOREST USERS

About 85 per cent of Bolivia's total forest area (534,000 square kilometres) lies below 500 metres (m) above sea level (MDSMA, 1995). Most of these lowland forests are east and north-east of the Andes, and they range from humid evergreen forest in the north to dry deciduous forest in the south (Montes De Oca, 1989). The species abundance for timber production is higher in deciduous forest; but the estimated logging potential volume is greater in the evergreen forest (Dauber et al, 1999).

During the first half of the 1990s, the state still held the rights over all forest areas in both public areas and private properties. About 20.7 million hectares (ha) of the Bolivian lowlands (76 million hectares) had been granted to forest companies through short- and medium-term contracts, although only 3 million hectares was actually harvested every year (Hunnisett, 1996, p7). That system concentrated forests in the hands of a few. After the new Forestry Law was issued in 1996, 89 timber companies reduced their concession areas to 5.8 million hectares, mainly because of the change from a volume-based to an area-based forest fee (US$1 per hectare). By 2002, the portion under forest concessions had declined even further to 5.3 million hectares as three forest concessions reverted to state control.

A portion of the areas that reverted to the state had overlapping rights with indigenous people's land claims and protected areas. Since the early 1990s, indigenous people have been granted a total area of 22 million hectares, with only 3 million hectares titled. Forest management plans approved by the Forest Superintendence (*Superintendencia Forestal*) cover a total of 540,000ha. Furthermore, the government has allocated 23 million hectares to large and medium-sized farms, and 3 million hectares were distributed to small-scale farmers or colonists, who represent an important source of timber when they convert forest to agricultural uses. By 2002, only 544,000 hectares within small- to large-scale private landholdings had a formal forest management plan; 2.2 million hectares was in municipal forest reserves, but only 492,000 hectares of these were assigned to 18 groups, of which 387,000 hectares had an approved forest management plan (Cronkleton and Albornoz, 2004). Thus, about 1 million hectares of forest is now under community forestry management – an important change.

Environmental concerns led the Bolivian government to declare about 15 million hectares as protected areas in the lowlands, although only a small proportion is effectively protected. About 71 million of the 76 million hectares in total lowlands has some type of property rights or conservation status. Nevertheless, this estimate does not account for overlapping or uncertain property rights, the latter often devoted to non-timber forest products extraction. Furthermore, there are no reliable data about how much area remains as public forest.

The mid 1990s witnessed changes in the ways in which forest resources were both allocated and used in the Bolivian lowlands. First, indigenous people's rights over their territories were formally recognized. Second, private landholders were granted rights over the forest resources located within their properties; hence, they were allowed to conduct logging activities. Finally, small-scale loggers, who were used to working informally within forest concessions and protected areas because they had no legal rights to access forested areas, also gained access to some forest resources within municipal forest reserves. These events have prompted the development of initiatives of community forestry management, leading to increasing cooperation among forest users.

Social participation and forest policy

Decentralization in Bolivia was motivated by increasing pressure from civic committees seeking to have greater control over natural resources, the general trend towards decentralization in neighbouring countries and the prominence of decentralization in donors' agendas (Kaimowitz et al, 1999). Its implementation has been affected by social participation dynamics, and the new forestry regulations have had implications for democratization of forest resources access.

Antecedents of decentralization

Bolivia has three levels of government: central government; departmental government or prefecture, whose main authority (the *prefecto*) is appointed by the president; and municipal government, in which democratically elected municipal councils elect the mayor. Before decentralization most decisions were made at the central level. The municipalities had limited resources and little influence in policy-related decisions, even those directly affecting the development of their municipal jurisdictions.

The Bolivian lowland forest area was largely marginalized from the political centre in La Paz, where most political decisions were made. During the early 1960s the region was progressively integrated within the national economy through the expansion of natural gas extraction, agriculture and logging. Those factors fostered the growth of a regional elite, expanded the contribution of the region to the national income and increased its influence on development programmes. During the 1970s, the government took its first step towards decentralization by establishing corporations for regional development. This was closer to an administrative attempt to transfer some investment decisions to the departmental level; but the central government still appointed the presidents of such entities, and most of the decisions were negotiated at higher levels (Kaimowitz et al, 1998).

Civic committees (groups of local social organizations) had been seeking greater access to forest revenues and greater participation in the formulation of forest policies since the mid 1970s. During the 1980s, legislation was approved establishing the collection of a forest fee of 11 per cent, to be used in regional development projects (in 1993, further legislation was passed stating that forest companies would pay 80 per cent of their taxes directly in the areas where the resources originated; but the mechanism was hard to implement in practice). By the late 1980s, the national forestry service was deconcentrated to local branches. This move towards deconcentration, however, did not make the forestry service a more efficient and less corrupt institution. Municipal authorities, meanwhile, continued to respond to the leaders of their respective political parties rather than to their constituents.

Decentralization was prompted through approval of the Popular Participation Law (No 1551) and the Administrative Decentralization Law (No 1654), both passed by congress in 1994. The first altered the responsibilities of the municipal governments, while the second modified the responsibilities of prefectures or departmental governments. In 1996 a new Agrarian Reform Law was issued, as well as a new Forestry Law. The Agrarian Reform Law's objective was to define the legal basis for a system of titling and land regularization, and to redefine the conditions of access to rural property. The Forestry Law attempted to redefine the conditions for obtaining and maintaining forest rights. Both affected the way in which landholders and forest users can access and maintain their rights for forest resources use.

Participation in investment decisions

The Popular Participation Law expanded the municipal government's jurisdiction beyond the urban centre to the whole rural area within the municipal borders. It made municipalities responsible for local schools, health facilities, roads maintenance and water systems. To finance these new responsibilities, the central government allocated 20 per cent of the national budget to the municipal governments, to be distributed among municipalities in proportion to their populations. Both rural and urban property taxes were earmarked for the municipal governments, who now administer their collection (SNPP, 1994).

The Popular Participation Law has strengthened municipal governments and made them more democratic. Rural populations – mainly smallholders and indigenous people – have gained the right to participate in municipal elections and run for the municipal councils. Nonetheless, national political parties still appoint individual candidates, allowing political parties to maintain their control over local political agendas and to reproduce a system of political patronage with local leaders. Local candidates who want to run for office have to negotiate their agendas within the priorities of the political parties, and if they are elected they are accountable to those parties.

Furthermore, the law sought to introduce community control over municipal governments by local social organizations (that is, local farmer organizations, neighbourhood committees and indigenous groups) and community-based vigilance committees (Rojas, 1996). Nevertheless, these committees' representatives are exposed to continuous pressures from political parties.

Rules and regulations for forest use

The Forestry Law of 1996 defined a set of regulations for forest use, somewhat differentiated according to forest user, under the premise that sustainable forest management is feasible under the right practices. It established a new system for monitoring forest management, enforcement and sanctions against illegal logging, as well as introducing some market-oriented regulations and taxes to discourage unsustainable forestry operations. The goal is to achieve sustainability of forest management through progressive incorporation of less valuable timber species and the application of extraction techniques that promote natural regeneration. Furthermore, the law seeks to create clear rights over forest resources, thereby encouraging investments in forest management; and to eliminate forest crime and illegal logging, as well as set technical criteria for forest management.

The public institutional system of the forest sector was substantially altered. The Ministry of Sustainable Development and Planning is the ruling entity, the Forestry Superintendence is the regulation entity, and the National Forestry Development Fund is the financial entity.

Non-commercial forest uses do not require authorization, and a forest management plan is an essential requirement for all types of commercial forest activities. Hence, forest concessionaires as well as private landholders are obligated to design management plans as an instrument to regulate commercial logging activities, including forest inventories and mapping. Forest management plans have to comply with many technical requirements. Forest management, when based on selective management, must respect a minimum cycle of 20 years between logging operations on the same area, and a minimum cut diameter must be respected. Furthermore, annual operations plans are required. The regime to be applied to non-timber forest products is similar (BOLFOR, 1997).

Democratizing access to forest resources

The new forestry regulations included two provisions that have had some impact in democratizing access to forest resources. The first refers to the exclusive right of indigenous peoples to use the forest resources within their indigenous territories, recognized legally by the Agrarian Reform Law. According to this law, indigenous claims are considered titled after completion of a process of land regularization. Currently, a total of 19 million hectares have been claimed for titling as indigenous territories; but it is not known how much will ultimately be effectively titled in this way. Stocks (1999) estimated 5 million hectares to be the area with commercial logging potential in areas claimed as indigenous territories.

A provision of the Forestry Law states that local forest user groups can benefit from forest concessions within areas declared as municipal forest reserves, which represent up to 20 per cent of public forest within each municipal jurisdiction. Local forest users can be granted a forest concession if they are recognized as a local user association by the Ministry of Sustainable Development and Planning. This mechanism was conceived as a way of formalizing the access for local forest users or small-scale loggers who, because they had no legal right to access forest resources, were previously conducting forestry operations informally (Kraljevic, 1996).

Powers transferred to municipal governments

According to Ribot (2001), outcomes of decentralization depend upon the type of powers that are transferred to lower levels and upon who receives such powers. In Bolivia, as well as in most countries of Latin America, decentralization has followed a top-down format, and municipalities have been the main recipients of authority transferred from the central level.

A top-down model of delegation

Although the Popular Participation Law did not grant municipal governments any new explicit function related to natural resources management, it motivated some municipal governments – as a result of their larger political authority – to become involved in natural resources issues and thereby capture part of the benefits. Because mayors became relatively more powerful, the central government and the international donors began to consider them more seriously as partners in environmental projects (Kaimowitz et al, 1999). The increasing political power of municipalities led the government to consider them for dealing with some problematic issues, such as social monitoring of illegal logging and formalizing forest rights for small-scale loggers and other local users. These were included in the Forestry Law of 1996.

The current powers and functions are divided as follows.

Ministry of Sustainable Development and Planning

This ministry:

- formulates forest policies, strategies and norms;
- determines land classification and evaluates forest potential;
- prepares demarcation of concession areas for timber companies and local groups;
- sets prices for concession fees and volume-based taxes;
- promotes research, extension and education; and
- solicits technical assistance and funding for forestry projects.

Forest Superintendence

The Forest Superintendence:

- supervises technical compliance with the forestry regime;
- grants management rights to eligible forest users;
- approves forest management plans for different forest rights;
- enforces forest regulations and sanctions illegal forest users;
- issues concessions, authorizations and logging permits;
- requests external forest audits of forest operations; and
- collects concession fees and volume-based taxes and distributes them.

Prefectures

Prefectures:

- formulate forest development departmental plans;
- develop forest research and extension programmes;
- promote programmes of rehabilitation of degraded forest systems; and
- develop programmes for strengthening municipal forestry units' institutional capacities.

Municipal governments

Municipal governments:

- propose the delimitation of municipal forest reserves up to 20 per cent of available public forest;
- protect and conserve the reserve areas until they are conceded to local user associations;
- inspect and control all forest activities within their territorial jurisdiction;
- report violations of forest regulations to the Forest Superintendence;
- provide support to local forest users in implementing their management plans;

- establish the registry of forest resources in their jurisdiction;
- develop soil use plans corresponding to the departmental use plans;
- organize training events for local forest users; and
- facilitate and promote social participation in local forest development.

In order to carry out their new responsibilities, municipal governments are expected to create municipal forestry units. Municipalities can form consortiums with other municipalities to create such units. In theory, the entire system should be entirely financed with the revenues coming from both concession and clear-cutting fees. Prefectures receive 35 per cent of the concession fees and 25 per cent of the fees charged for clear-cutting operations. Municipal governments get 25 per cent of both types of fees. The National Forestry Development Fund receives 10 per cent of the concession fees and 50 per cent of the clear-cutting fees. The Forest Superintendence gets 30 per cent of the concession fees. A reduction of forest concession fees in March 2003 has diminished the financial resources going to prefectures and municipalities.

Restricted powers for municipalities

Municipal governments must comply with national regulations regarding property regimes and forest regulations. As a result, their autonomy to make decisions about natural resources depends largely upon decisions made at the central level. The legislation thus regards municipal officials as rule followers rather than rule makers, and as implementing agencies for policies defined at the central level.

In short, the new institutional forest system has not delegated important responsibilities to municipalities and has therefore not led to dramatic changes. The central level reserves major decisions regarding allocation of forest resources rights, granting forest concessions, approving forest use regulations, and collecting taxes from forest resources use for itself. Municipalities have little to say in these areas, and all informal actions of municipalities for collecting forest taxes and controlling timber transit are considered illegal.

Although municipal governments are empowered to control forest crime, promote community forestry and take other forest-related actions, their power to make autonomous decisions regarding the allocation and use of forest resources is restricted. For instance, they can decide how to allocate the forest resources within the municipal reserves, but cannot decide about the size of such reserves. This reflects the tension between the central level, which defends the strong role of the national forestry service, and those who support a more active and autonomous role for municipal governments.

Limited autonomy for local forest users

The situation of indigenous groups has improved. Indigenous communities whose forest resources were subject to encroachment from illegal loggers now have exclusive access to resources within their territories, although problems persist (Roper, 2000). However, before indigenous people can take commercial advantage of forest resources, they have to develop forest management plans according to Forest Superintendence regulations, and they have little scope to adapt such norms to their own management practices (BOLFOR, MDSMA, and Superintendencia Forestal, 1997). And even though specific regulations for forest management in indigenous areas have been approved, local knowledge has been ignored in much of the forest policy. Enforcement of indigenous property rights depends upon some institutional arrangements made at the departmental and regional levels over which indigenous people have little influence. This issue is linked to a more complex bureaucratic process of land-use planning and titling with poor overall outcomes.

The other social actors who have benefited from access to forest resources – small-scale timber extractors and other local forest users – also have no autonomy to make decisions about the way in which they use the resources, and all of them have to comply with forestry regulations to maintain access to public forest resources through forest concession systems. The Forest Superintendence has instituted command-and-control mechanisms to enforce the implementation of what it considers good forest management practices among local forest user groups and small landholders. Those practices, though appropriate for large-scale forest concessionaires, act as barriers for small-scale forest users (Pacheco, 2001).

Implementation of decentralization policies

There are always failures in implementing policies, and decentralization is no exception. In Bolivia, transferring responsibilities to municipalities meant building local capacities in municipal governments, as well as interacting with actors who had disparate incentives and interests to use the forest within the new institutional context. This section reviews such situations and discusses how closely implementation matched the original plans.

Building institutional capacities

Municipalities with forest resources began to receive their shares of forest taxes in 1997, when they started to set up municipal forestry units. The resources transferred to municipalities have tended to decrease because forest concessionaires did not comply with the forest fees payments. In 2002, a new system formally reduced the amount of taxes collected by the state from forest operations to almost half of what was originally expected, formalizing the reduction of transfers from the central government to municipalities.

By the end of 2001, almost all municipalities with forest resources (about 109) had created their municipal forestry units, and each had at least one forest or agricultural technician. In municipalities with difficult access and little population, the forest units are the only local providers of technical services, and governments tend to value their role in local planning. The main limitations are the staffs' lack of technical skills and inadequate training in social issues, such as conflict resolution (Andersson, 2002).

Building technical capacities in municipal forestry units constituted an important step forward in relation to the past. Nevertheless, progress is not uniform and varies with the financial support the units receive from the municipalities. This, in turn, depends upon the amount collected in forest fees and what proportion of it goes to supporting operational budgets. A significant portion of municipalities consider it enough to provide minimum resources to the municipal forestry units and spend the rest, if any, on other activities. It is worth mentioning that local government priorities are providing social services and infrastructure, and they have little motivation to support productive projects (Faguet, 2001), although this is changing.

Incentives for the actors

The new institutional system resulting from both decentralization and the shift in forest policies has modified the incentives for municipalities and forest stakeholders to continue doing what they were used to doing, or to adapt their social and financial strategies to the new conditions. The main incentive for municipalities to engage with the new decentralized order of things was the likelihood of getting a share of the forest taxes, as well as the possibility of administering the municipal forest reserves. Municipalities are also interested in penalizing illegal clear-cutting due to the fact that they get a portion of the fines; but this does not happen in relation to illegal logging. Furthermore, municipal governments have been interested in being active players on forest issues in the cases where local forest users have political influence on local decision-making, or in cases where forest-dependent people represent important votes to keep them in office.

The incentives of local forest users to engage in the process are diverse. Indigenous people have received important benefits from forest policy reform and decentralization. They have gained rights to make exclusive use of their forest resources within their territories, and have the chance to expand their influence to participate in municipal decision-making. In this context, forest management might increase the social legitimacy of indigenous people's claims to land. In turn, local small-scale timber extractors' main incentive has been to get formal access to forest resources through the forest concession system, and to benefit more from formal markets. Furthermore, this group should benefit from the technical assistance provided by the municipal forestry units, which, as mentioned earlier, are receiving resources from the forest taxes to support such activities. Both groups have benefited from several forest projects (for example, the project for Sustainable Forest Management, BOLFOR) and non-governmental organizations (NGOs) with an interest in supporting community forestry.

In the short run, the main losers under decentralization were absentee forest concessionaires, who must now negotiate with the forestry service. Furthermore, a forest tax has been imposed, and they have had to acknowledge indigenous people's demands. The main incentive for forest concessionaires to adapt to the new conditions was their interest in keeping their forest areas, and an implicit commitment from the central government to help them through a difficult financial situation originating in the timber crisis in regional markets. Municipal governments reacted in contradictory ways regarding forest concessions. Although some are critical of concessions' activities within their jurisdictions, others consider them sources of economic growth and employment (Pacheco and Kaimowitz, 1998).

Homogeneous solutions for heterogeneous municipalities

The Bolivian model of decentralization does not account for regional variation, and its design assumes that all municipal governments will react uniformly to the challenges arising from the new conditions. Reality proved rather different. Implementation of decentralization has had to face three issues: uneven distribution of resources to municipalities; the minimal availability of public forest declared as municipal forest reserves; and the heterogeneous interests of municipalities regarding forest-related activities.

Uneven distribution and allocation of financial resources

The financial resources allocated by the Forest Superintendence to municipalities have varied. From 1997 to 1999, only 30 municipalities benefited, with more than 80 per cent of the total resources transferred to them. In the richer municipalities, not all income from forest taxes is spent on forest-related activities; some is diverted to other sectors. The poorest municipalities likewise have urgent demands that are outside the forestry sector (de Urioste, 2000; Flores and Ridder, 2000).

Minimal municipal forest reserves

Even though the law specified that up to 20 per cent of public forest would be declared municipal forest reserves, in practice such areas were not available in all municipalities – in some cases because of the existence of overlapping claims over public forest, and in others because public lands were maintained as forest concessions. In some municipalities there were no areas to declare as municipal reserves to allocate as concessions to small-scale loggers. By early 2001, from a total of 2.4 million hectares demanded by municipalities as municipal reserves, only 681,000 hectares had been designated (Pacheco, 2003).

Municipalities' priorities

Only a few municipal forestry units have accomplished the functions that they have been granted. Municipalities prioritize their investment in response to social pressures or to local authorities' political motivations. The level of municipal engagement is also related to municipalities' incentives to respond to their constituents and to be accountable to higher authorities (Andersson, 2003). For instance, whereas some municipalities are interested in allocating resources to establish a system of forest concessions within municipal reserves, others choose to control illegal clear-cutting or make forest management viable for small farmers.

UNEVEN OUTCOMES

Transferring responsibilities to municipalities has modified the political and institutional arrangements for forest management, with implications for both people and forests.

Participation in local politics

In many lowland municipalities, small farmers, indigenous people and small-scale loggers have been elected to office for the first time. Indigenous groups have been able to obtain political support from the municipal councils to reinforce their land claims, small-scale loggers have obtained support to negotiate temporary logging authorizations, and small farmers have obtained support favouring their efforts to modify land use and forest regulations. Thus, municipal governments have contributed to political support for some local actors' claims on resources. This is also the case where these groups have strong organizations that can influence the municipality's decisions or where they represent the majority of voters. In these cases, municipal governments may amplify the demands of social actors.

In other cases, however, transferring responsibilities and resources to municipalities has reinforced the power of pre-existing local elites – elite capture, particularly in municipalities of northern Bolivia, and where cattle ranchers and timber companies are highly influential in local politics. These local elites have, in some cases, influenced municipal governments to build local alliances against indigenous land claims and strengthened their power over use of resources. The local elites of these municipalities often benefit from activities based on mining of natural resources and promote concentration of benefits in a few hands. Decentralization, in these cases, tends to produce undesired results in social equity and forest conservation.

Where the social composition of local government is more complex, local elites have to negotiate with small farmers and indigenous people (Pacheco and Kaimowitz, 1998). It is not unusual to find alliances built among the different social groups that support certain development agendas, such as infrastructure development and basic social services. Those social agreements are, however, difficult to arrange for natural resources management, although in some cases alliances have been established to protect conservation areas against encroachment of foreign timber and mining companies (that is, municipalities of Rurrenabaque and San Ignacio de Velasco).

Much of the impact of decentralization on reconfiguring social participation in local politics depends upon the local political economy and upon the social capital of marginalized groups. In other words, the more democratic municipalities have traditionally had more equitable access to resources. Those in which local elites have imposed their interests are affected by the extent to which local groups have been able to build social capital.

Rights of forest-dependent people

One of the most striking features of decentralization in Bolivia is that the new, more diverse social composition of municipal governments has had little effect on how forest rights are allocated and to whom: such decisions remain at the central level. In this regard, municipal governments become institutions for stimulating negotiations among local actors and representing local demands to the national level. Indeed, municipal governments were active players in making some forest regulations more flexible, which benefited small-scale loggers and small landholders.

The reinforcement of some property rights that accompanied decentralization, such as indigenous territories and municipal forest reserves, had significant implications for local forest users, though the pace of such changes was extremely slow. Titling of indigenous territories is a slow, bureaucratic process, and indigenous people cannot benefit fully from their forest resources because they do not fulfill the conditions to formulate and implement forest management plans (Stocks, 1999).

Forest concessions for small-scale loggers have also proceeded slowly. Identification and demarcation of municipal forest reserves by the municipalities was relatively quick; but the approval of such areas by the Ministry of Sustainable Development and Planning was slow. Besides bureaucratic procedures, one of the factors limiting the formal creation of municipal reserves was an extremely slow process identifying public forest, which is a precondition for establishing such reserves (Pacheco, 2000). Many of the claimed areas are not available. Moreover, about 26 demands of local user associations were not even processed, leaving the door open for such groups to persist in illegal logging activities.

The impact of municipal forest reserves is still small: only 13 groups with 387,000ha had an approved forest management plan by early 2003. Unfortunately, there are no assessments of the extent to which such groups have improved their incomes from their logging operations. Members of the local user associations, lacking any tradition of collaboration because they were accustomed to working independently, have faced internal conflicts of leadership (Cronkleton and Albornoz, 2004). Furthermore, these groups' organizational procedures and mechanisms for distribution of profits from logging operations remain unclear (Kraljevic, 2002). Building social capital in these groups might be more important in the long term than improving their practices of forest management.

Impacts on forest crime

Many changes in forest management practices can be related to the shift of forest policy. Indeed, the Forest Superintendence has been more active in monitoring forest crime, both illegal logging and clear-cutting, though it was not able to slow illegal activities because of its limited institutional capacity to enforce forestry regulations. There are no reliable estimates of illegal logging, and those of clear-cutting suggest that illegal deforestation would be around 80 per cent of total deforestation (Contreras, 2001). At first, monitoring activities focused on enforcing small landholders' and illegal forest users' compliance with the forest regulations (Pacheco, 2001). Thereafter they focused on municipalities with high rates of informal activities.

Although the Forest Superintendence carried out most monitoring activities after decentralization, gradually it sought help from the municipal forestry units – both to use the resources transferred to municipalities to develop monitoring activities, and to justify its actions before the local population. Nevertheless, municipalities' response to forest crime monitoring was ambiguous. Controlling illegal forest activities would affect some politically influential people within some municipal governments, and units had no capacity or resources to spend on activities with little financial return and high political cost. That institutional behaviour has created ambiguous signals about the role that municipalities play in forest crime monitoring.

Municipal governments have more incentive to control the operations of large-scale forest concessions and to control illegal clear-cutting because of the direct benefits they obtain. A few municipalities have taken over machinery of timber companies and intervened when forest concessionaires logged in areas outside their boundaries. Local authorities complain about the system of auctioning confiscated illegal timber because they do not benefit from it. They have been active in temporarily resolving uncertainty over property rights, primarily of small landholders, by issuing land 'possession certificates', although these lack legal value.

Institutional tensions and collaboration

The municipalities and the Forest Superintendence are beginning to build partnerships to improve forest management monitoring at the local level. This agency was initially sceptical about local authorities' ability to intervene in forest-related issues in a politically neutral way. The municipal units view it as undermining the livelihoods of local forest users who have undergone severe difficulties to adapt their practices to the new forestry regulations.

That has changed to some degree, and the Forest Superintendence is now more involved in building collaborative agreements with municipal forestry units for regulations enforcement. The driving principle is the need to enhance local operational capacity and to make legitimate decisions at the local level. Through formal agreements of collaboration, the agency delegates some responsibilities to certain municipal forestry units. In other cases, however, there is a lack of communication and trust among the personnel involved and some previous negative experiences (Flores and Ridder, 2000).

NGOs have helped municipal forestry units to overcome institutional deficiencies and build capacity for monitoring, planning and proposal elaboration. Some forest projects – such as the BOLFOR project working with local user associations, and the United Nations International Drug Control Program (UNDCP)–Food and Agriculture Organization (FAO) project working with colonists – attempted to build capacity for forest management through training in preparing and implementing forest management plans, developing market skills and, in some cases, establishing partnerships with timber companies. In most cases, however, forest projects have not included municipal governments working directly with forest users. That might have had some positive effects in such projects' achievement of their goals; but inclusion of municipal units would have contributed better to the effort to develop technical skills within municipalities.

The missing link in the system is at the level of the departmental governments. Even though they receive resources from a portion of the forest taxes, they have resisted spending them on forest research and extension and on strengthening municipal forestry units. Instead, those resources were diverted to cover other expenses. An important exception is an isolated project aimed at strengthening municipal forestry units developed by the Prefecture of Santa Cruz between 1998 and 2000, which has been documented elsewhere (Flores and Ridder, 2000). There is an ongoing debate over the role of prefectures within the forest institutional system.

The prerequisites of success

Decentralization contributes to the success of municipal forest management in cases where:

- It promotes more equitable access to forest resources for local forest users who depend, to some extent, upon forest resources,
- It stimulates the growth of financial resources invested in forestry initiatives, with multiplier effects in the local economies.
- It facilitates the development of actions to plan better management of forest resources and natural resources conservation.

Yet, all of these conditions are difficult to achieve simultaneously in practice, and trade-offs are inevitable.

Decentralization has had positive outcomes on equity because it has expanded the influence of previously marginalized social groups (mainly indigenous people and small landholders) in decision-making. Rural people gained the right to run for office and began to make use of the mayors' mechanisms of social control. This did not happen everywhere, however. The factors that helped to extend social participation in decision-making were strongly related to the pre-existence of social networks that enabled collective action and kept municipal authorities accountable to local social organizations (for example, indigenous municipalities and municipalities dominated by coca growers in Chapare). Such situations are more likely where there is more democratic access to land, and where institutional rules ensure rights of access to forest resources. Decentralization has, broadly speaking, strengthened indigenous claims over land and allowed for the creation of municipal reserves.

The impact that greater equity has on local development is not yet clear. Most decisions on the allocation of forest rights lie outside the competence of municipal governments, which have little voice about forest concessions allocation within municipal forest reserves. Furthermore, the increasing influence of marginalized groups on municipal government decision-making does not automatically lead to influence over allocation decisions, although it often enables them to pressure the central government. In general, municipalities where local grassroots organizations affect municipal decisions are more empathetic to the recognition of forest rights for local user groups. Secure property rights have undoubtedly improved the livelihoods of small-scale loggers and indigenous people. Yet, in some cases, forest regulations have made life difficult for these same groups, and making forest regulations more flexible has been a slow process.

Expanding local control over monitoring of forest resources use has a two-sided effect. On the one hand, forest crime continues where local politicians are part of corruption schemes. On the other hand, local populations have more incentive to protect natural resources from absentee landholders or companies without local roots. A similar logic applies to protected areas: they are resisted when they may negatively affect local people's livelihoods or local elite interests, but accepted when they may bring some resources to municipalities. Reinforcing property rights and increasing mechanisms of social control have, in some cases, contributed to making municipal decisions on monitoring more politically neutral.

In short, the Bolivian experience shows that the top-down move towards decentralizing forest management is accompanied by social forces emerging from the bottom. Different political economic contexts and local power relationships have led to diverse outcomes of decentralization for improving livelihoods of local forest users and sustainable forest management. Nevertheless, by redefining local democracy, decentralization has created new conditions under which local actors can participate in politics and influence territorial local development. The power of municipalities over forest resources is still limited; but local capacity development may change that situation.

By becoming more democratic, decentralization in Bolivia will enable municipal governments to exert more power over the allocation of forest resources, make property rights more secure, penalize violators with more certainty, and give local forest users' institutions more rights to manage and use their resources. A better balance in the distribution of competences has to be built among the central government, municipal governments and local institutions, accounting for both local variations and the disparate incentives and interests of the local actors. Three issues demand greater attention:

1 building horizontal systems of monitoring in which trust is at the core of collaborative actions in order to enhance social capital of local forest users;
2 helping municipal governments to become more transparent and participatory; and
3 developing pathways of improving the contribution of forests to poverty alleviation and local development rooted in local capacities and capabilities.

Although decentralization is only one of the political and economic forces that contribute to shaping the relationship between forest and livelihoods, strengthening decentralization can make a difference.

References

Andersson, K. (2002) *Explaining the Mixed Success of Municipal Governance of Forest Resources in Bolivia: Overcoming Local Information Barriers.* Natural Management Resources, World Bank, Washington, DC

Andersson, K. (2003) 'What motivates municipal governments? Uncovering the institutional incentives for municipal governance of forest resources in Bolivia'. *Journal of Environment and Development,* vol 12, no 1, pp5–27

BOLFOR (1997) *Ley y Reglamento de la Ley Forestal, Ley INRA.* Proyecto BOLFOR, Santa Cruz, Bolivia

BOLFOR, MDSMA, and Superintendencia Forestal (1997) *Normas Técnicas para la Elaboración de Instrumentos de Manejo Forestal Comercial (Inventarios, Planes de Manejo, Planes Operativos, Mapas) en Tierras Comunitarias de Origen (Resolución Ministerial No 136/97).* Santa Cruz, Bolivia

Contreras, A. (2001) *Forest Law Compliance: An Overview.* World Bank, Washington, DC

Cronkleton, P. and Albornoz, M. A. (2004) *The Diversity of Community Forestry in Bolivia.* Center for International Forestry Research (CIFOR), Santa Cruz

Dauber, E. R., Guzmán, A. and Terán, J. R. (1999) *Potencial de los Bosques Naturales de Bolivia para Producción Forestal Permanente.* Superintendencia Forestal, Santa Cruz, Bolivia

de Urioste, J. L. (2000) *Informe Final.* Proyecto de Apoyo a la Gestión Forestal Municipal, Superintendencia Forestal/Cooperación del Gobierno de Canadá, Santa Cruz, Bolivia

Faguet, J. F. (2001) *Does Decentralization Increase Responsiveness to Local Needs?*

Evidence from Bolivia. World Bank, Washington, DC
Flores, G. and Ridder, M. (2000) *Experiencias con el Proceso de Fortalecimiento Forestal Municipal en Santa Cruz*. FAO-PAFBOL, Santa Cruz, Bolivia
Hunnisett, G. (1996) *The Forest Sector and Deforestation in Bolivia*. World Bank, La Paz, Bolivia
Kaimowitz, D., Vallejos, C., Pacheco, P. and Lopez, R. (1998) 'Municipal governments and forest management in lowland Bolivia'. *Journal of Environment and Development*, vol 7, no 1, p45
Kaimowitz, D., Pacheco, P., Johnson, J., Pavez, I., Vallejos, C. and Vélez, R. (1999) 'Local governments and forests in the Bolivian lowlands'. *Rural Development Forestry Papers* (RDFN) No 24b, Overseas Development Institute, London
Kraljevic, I. (1996) *Estudio Exploratorio del Sector Maderero Local de la Provincia Velasco en el Departamento de Santa Cruz*. Documento Técnico 48/1996, Proyecto de Manejo Forestal Sostenible BOLFOR, Santa Cruz, Bolivia
Kraljevic, I. (2002) *Evaluacion Rapida de Temas Organizativos que Afectan al Desarrollo de las ASLs*. Documento tecnico 109/2002, BOLFOR, Santa Cruz, Bolivia
Larson, A. (2003) 'Decentralization and forest management in Latin America: Toward a working model'. *Public Administration and Development*, vol 23, no 2, pp211–226
MDSMA (1995) *Memoria Explicativa. Mapa Forestal*. Ministerio de Desarrollo Sostenible y Medio Ambiente, Secretaría Nacional de Recursos Naturales, La Paz
Montes De Oca, I. (1989) *Geografía y Recursos Naturales De Bolivia*. Ministerio de Educación y Cultura, La Paz
Pacheco, P. (2000) *Avances y Desafíos en la Descentralización de la Gestión de los Recursos Forestales en Bolivia*. CIFOR/BOLFOR, Santa Cruz, Bolivia
Pacheco, P. (2001) *Bolivia: Country Profile for the Forum: The Role of Forest in Poverty Alleviation*. Forestry Department, FAO, Rome, Italy
Pacheco, P. (2003) 'Municipalidades y Participación Local en la Gestión Forestal en Bolivia', in Ferroukhi, L. (ed) *La Gestión Forestal Municipal en América Latina*. Center for International Forestry Research, International Development Research Center, Bogor, Indonesia
Pacheco, P. and Kaimowitz, D. (1998) *Municipios y Gestión Forestal en el Trópico Boliviano*. Center for International Forestry Research, Centro de Estudios para el Desarrollo Laboral y Agrario, Fundación TIERRA, Proyecto BOLFOR, La Paz
Ribot, J. C. (2001) 'Local actors, powers and accountability in African decentralization: A review of issues'. Paper presented at Decentralization and Local Governance in Africa, 26–30 March 2001, Cape Town, South Africa
Rojas, G. (1996) *La Participación Popular: Avances y Obstáculos*. Secretaría de Participación Popular, La Paz
Roper, J. M. (2000) 'Whose territory is it? Resource contestation and organizational chaos in Bolivia's multiethnic indigenous territory'. Paper presented at the Latin American Studies Association 2000, Miami, Florida
SNPP (1994) *Ley de Participación Popular No. 1551*. Secretaría Nacional de Participación Popular, La Paz
Stocks, A. (1999) *Iniciativas Forestales Indígenas en el Trópico Boliviano: Realidades y Opciones*. Documento Técnico 78/1999, Project for Sustainable Forest Management BOLFOR, Santa Cruz, Bolivia

Chapter 10

Decentralization and Forest Management in Uganda

Steve Amooti Nsita

INTRODUCTION

Local governments in Uganda manage 5000 hectares (ha) of local forest reserves and oversee the forests on private lands, which constitute 70 per cent of Uganda's forests. Ten years after the decentralization process was introduced in Uganda, it has not really embraced forests. The problem is that it was not tried out long enough to generate confidence on both sides (central government and local governments) and to allow time for the local governments to build their capacity. Friction between the central government's Forestry Department and local governments continues because the laws governing local governments and those governing forestry have not clarified whether forests are decentralized or not. The ensuing confusion has led to deterioration of forests through illegal harvesting and encroachment for agriculture and settlement. Local forest reserves created in 1998 were not transferred to local governments with the corresponding resources to manage them. As a result, local governments did not take up the responsibility. These reserves also became heavily encroached. Decentralization has inspired almost no public or private investment at the forest level. Nevertheless, definable individual financial benefits indicate the beginnings of private capital investment in forestry.

FOREST MANAGEMENT IN UGANDA

Uganda (population 24.7 million people) covers a total surface area of 241,038 square kilometres. Subsistence farmland covers 41 per cent of the land area; forests cover 24 per cent; and bush land covers 7 per cent. The remaining 28 per cent comprises grasslands, water, marshland and built areas (Ministry of Water, Lands and Environment, 2001). Forests cover 4.9 million hectares and consist primarily (81 per cent) of woodlands (4m in height but less than 30 per cent canopy cover), with tropical high forests (19 per cent) and plantations (less than 1 per cent).

Thirty per cent of the forests are managed by government agencies (the Forestry Department, which has been reconfigured as the National Forestry Authority), local governments and the Uganda Wildlife Authority. The permanent forest estate is composed of 1.9 million hectares, representing about 9 per cent of the total land area of Uganda (Ministry of Water, Lands and Environment, 2001). It covers all forest reserves (1.2 million hectares) and forested areas in national parks (0.7 million hectares). Seventy per cent of the forests are to be found on private lands.

Forestry agency reports show that 50 per cent of the tropical high forests on private lands are degraded, and 17 per cent of those in protected areas are degraded. The main reasons for the degradation include harvesting for timber, firewood and charcoal, and encroachment for agriculture and human settlement. Forest industries are small-scale and owned almost entirely by Ugandans. Markets are also largely domestic, although there are increasing exports of value-added forest products, especially within the region. It has become apparent that efforts to promote responsible forest management should be directed at owners of private lands with forests.

Historical background

Forest management in Uganda has been decentralized and recentralized a number of times since independence in 1962 as different governments adopted varying policies. Until the late 1990s, forest management in Uganda was mainly a public matter, more or less confined to forest reserves.

Before 1967, there was a vibrant local forestry service, which ran local forest reserves, especially under the kingdoms that had built strong, coherent government systems. At that time, local governments had powers to decide on development priorities for their areas. Nevertheless, the central government was still responsible for managing some of the forest reserves.

In 1967, the Ugandan government adopted a republican constitution, which centralized virtually all government decision-making powers. Consequently, all local forest reserves became central forest reserves.

In 1993, the government adopted the policy of decentralization. However, it soon became clear that local governments were not yet ready to assume all forest management responsibilities. Needing revenue to run more urgent activities such as education, water and health, which had also been decentralized, they set about harvesting the forests with little consideration for planned management.

The forest reserves were therefore recentralized in 1995, but this time through subsidiary legislation rather than the principal law. This turned out to be a rather unpopular move. Local governments challenged the legal basis (albeit outside the law courts) for recentralizing the forest reserves and maintained pressure on the central government to decentralize them again.

In 1995, Uganda adopted a new constitution. The constitution fully embraced the 1993 decentralization policy; but it remained ambivalent regarding management of forests.

Since then the local governments have been asking the central government to hand over all forest reserves to them, arguing (probably correctly) that the law vested only policy in the central government. The Forestry Department replied that the local governments did not have the capacity and sufficient will to manage the reserves professionally.

The National Forestry and Tree Planting Act of 2003 maintained the 1998 state of affairs; but this time it created the semi-autonomous National Forestry Authority. Whereas the Forestry Department had been responsible for all aspects of forestry in the country, the new agency would manage only the central forest reserves. Indications are that local governments may go along with this approach.

The results of these changes have been mixed. Following independence, the local forest reserves were run efficiently, as were all other government activities. Up to the mid 1970s, the forest reserves were well managed even though they were centralized. Thereafter, efficiency depended upon available resources from the central government. The forest reserves have never been decentralized long enough to permit judgements about the impact of decentralization on sustainable forest management. Other decentralized sectors – notably health, education and roads – still depend upon grants from the central government, which maintains a strong influence on how things are run at the local level. Forestry is not likely to be different until local governments have built sufficient revenue bases of their own.

Decentralization in Uganda

The most widely understood form of decentralization in Uganda is *devolution* of financial and decision-making powers to sub-national structures at various levels. For rural areas, the hierarchy of local governments descends from district (the main sub-national administrative level) to county, sub-county, parish and, finally, village. In urban areas, the levels are city, city division, municipal council, town council, ward and village.

Article 176(b) of the Constitution of the Republic of Uganda, 1995, states:

> *... decentralization shall be a principle applying to all levels of local government and, in particular, from higher to lower local government units to ensure people's participation and democratic control in decision-making.*

Consequently, the purpose of the Local Governments Act, 1997, is as follows:

> *... to give effect to the decentralization and devolution of functions, powers and services; to provide for decentralization at all levels of local*

> *governments; to ensure good governance and democratic participation in, and control of, decision-making by the people.*

Although *deconcentration* transfers most of the day-to-day activities from the central government, all major decisions are still made there. The Forestry Department has been operating along those lines for a long time. For example, district forest officers undertake most operational responsibilities, such as forest management planning, budgeting and supervision of forest-level fieldwork; but the forest management plans and budgets have to be approved by the head of the Forestry Department in Kampala.

The general objectives of decentralization in Uganda are as follows (Kato, 1997):

- to transfer real power to the districts and thus reduce the load on remote and under-resourced central government officials;
- to bring political and administrative control over services to the point where they are actually delivered, and thus reduce competition for power at the centre and improve accountability and effectiveness;
- to free local managers from central constraints and thus allow them to develop organizational structures tailored to local circumstances;
- to improve financial accountability and responsibility by establishing a clear link between payment of taxes and the provision of the services they finance;
- to restructure government machinery to make the administration of the country more effective; and
- to create a democracy that brings about more efficiency and productivity in the state machinery by involving people at all levels.

Legal developments shaping decentralization

Until the late 1990s, management of forests was largely a function of the central government. Planned management has been almost entirely practised only in the forest reserves. There was virtually no planned management of forests on private lands until private companies and individuals started to acquire permits to grow their own forest plantations in grassland forest reserves during the mid 1990s.

Before 1967, when the government abolished the federal constitution written at independence, most of the forest reserves were managed through decentralized mechanisms. The mechanisms were mainly the *deconcentration* type, in which district officials did most of the work without central oversight, provided the annual plans and budgets were approved and adhered to. Financial allocations were conducted through departmental warrants from the central government after the plans and budgets had been approved. Barring significant variations in the annual work plan, districts usually operated more or less independently. At that time there were vibrant local governments, which had built coherent systems from the colonial days: they planned and executed all the activities in the districts, collected and disbursed revenue, hired and fired their own staff, and even had their own salary structure, different from that of civil servants in the central government.

In 1967, the government adopted a republican constitution, which centralized virtually all government decision-making powers. All forest reserves were now managed by the Forestry Department, and the local governments were stripped of all decision-making powers in matters of forest management, including forests on public lands.

At this time, Uganda had one of the best forest management services in Africa. Although management was centralized, the forest reserves were well run. Things started going wrong when the government was overthrown by the military in 1971.

In 1993, the government fully embarked on the decentralization process through the Local Governments (Resistance Councils) Statute, 1993. This statute decentralized forests. However, it soon became apparent that the local governments were not prepared to engage in professional forest management. They saw forests as sources of revenue to fund development activities in other sectors such as health, education and roads that had been decentralized (see Indonesia in Chapter 7).

As a result, forests were recentralized in December 1995 under Statutory Instrument No 52. By now, local officials had become fully aware of their powers, and the electorate valued its ability to influence government decisions. People began encroaching on forest reserves with the tacit consent of their elected leaders, and illegal harvesters openly dealt in forest products (especially timber and charcoal). Soon, some of the Forestry Department staff, finding themselves working against political forces that obliged their removal from office, quietly joined the elected leaders to collude with encroachers and illegal forest products dealers.

Since then, the Forestry Department has never managed to reverse this state of affairs. In fact, most of the serious cases of encroachment and illegal harvesting in Uganda today started or expanded considerably during the election periods. The intractable encroachment problems in South Busoga and Luwunga reserves grew. For example, even after the government had won a court case against the encroachers in Luwunga, the local political leaders made it impossible for court orders to be carried out; thus, decentralized powers caused a miscarriage of justice.

That is probably why the constitution of 1995 is ambivalent regarding management of forests. One clause authorizes 'Government or local government as determined by parliament by law' to hold forests in trust for the people of Uganda. Another leaves forest and game reserve policy and the 'environment' with the central government, but without mentioning forests. Thus, forests are among 'any other functions and services not specified', whose management then falls to the district councils.

The legislators failed to take a decisive constitutional stand. They understood the need for forest protection; but they also could not openly stand against the people who had elected them. The constitutional process failed to resolve the social aspects of sustainable forest management.

Because the local governments never really accepted the recentralization that took place just before the constitution was promulgated, the 1997 Local Government Act specifically mentions forests as a function of local governments. This effectively decentralized the forests again. The government then created central and local forest reserves in an attempt to appease local governments and their tacit supporters at the central level. But the areas involved were small: the 192 local reserves totaled just under 5000ha, compared with 542 central reserves totaling 1,455,130ha. Thus, the impasse was not resolved and the social standoff between staff and people living near the forest reserves continued.

In July 1993, the Traditional Rulers (Restitution of Assets and Properties) Statute would have paved the way for the traditional rulers to reclaim the forests that belonged to their kingdoms before 1967. But the traditional rulers do not have legal authority to hold forest reserves because the Local Government Act does not legally define traditional institutions as local governments. Consequently, in 2001, an executive order under a memorandum of understanding between the central government and authorities of Bunyoro Kingdom providing for the return of some forest reserves to the king could not be implemented. Although the head of the Forestry Department gave the kingdom authorities written permission to take over management of a reserve, the government's solicitor general advised that the kingdom could not hold any reserves in trust for the people of Uganda. However, the Forestry Department did not have sufficient political backing to reclaim the reserve. Taking advantage of the impasse, the kingdom simply continued timber harvesting in an uncontrolled manner, resulting in the degradation of the forest plantation to the financial advantage of a few individuals (see Chapter 4).

Not even the 2001 Forestry Policy was decisive. For example, it says:

> *...efforts will be made to clarify the role of local governments in management of forest resources... Any ambiguities and contradictions in the provisions of the Constitution, the Local Governments Act and the Land Act with respect to the role of districts in forest sector development will be addressed.*

The 2003 National Forests and Tree Planting Act attempted to clarify these ambiguities by distinguishing among central forest reserves (to be managed by the National Forestry Authority), local reserves and community forests (to be managed by local governments) and private forests. Now the local governments can directly collect and use revenue that used to go to the consolidated account of the central government.

It is still too early to tell how the new policy and law will affect forest management. However, the parties involved seem to have been satisfied with these arrangements. Heated discussions will arise when it comes to apportioning the reserves between the local governments and the National

Forestry Authority; but when that is agreed, all parties may start to cooperate on the path to sustainable forest management.

Other legal and institutional frameworks

Land Act, 1998

Deriving from Article 237 of the 1995 constitution, the Land Act vests ownership of land in the citizens of Uganda; but, as stated earlier, the central and local governments can hold forest reserves and other natural resources on behalf of all Ugandans. Local officials may ask the central government to manage any of their resources; however, so far, none have formally invoked this constitutional provision.

In exercising their right to manage these resources, the central and local governments 'shall not lease out or otherwise alienate any natural resource' that they hold in trust for the people. However, they may grant licences, concessions or permits for use of the resources.

The act empowers people to use the land they own in any way subject to other existing laws. In terms of forestry, the laws most commonly referred to are those dealing with forests, the environment and wildlife. People can also access forest reserves for user rights; but the reserves cannot be degazetted without approval of parliament.

The act establishes land boards at the district level and land committees at parish levels to deal with transfer of ownership, conflict resolution, allocation of land not owned by any body and review of compensation rates, among other issues. Most of the powers and responsibilities formerly exercised at the center have now been decentralized.

Because of unanticipated financial implications, implementation of some aspects was delayed. Therefore, it is still too early to gauge the impact of the act upon sustainable forest management. However, what is clear is that many municipal and town councils are putting considerable pressure on the Forestry Department to degazette reserves in urban areas. The law allows this, but requires approval by parliament. No local government has thus far asked parliament for approval; but they have tacitly allowed the urban reserves to be developed: built areas are more desirable than open lands in urban areas.

Environment Management Statute, 1995

The Environment Management Statute creates the National Environment Management Authority and makes it responsible for ensuring that activities are conducted in an environmentally friendly manner. It provides for the establishment of local government environment committees to coordinate activities at various local levels.

The statute provides for voluntary tree growing for environmental purposes by landowners. However, if an environment committee considers an area at risk of environmental degradation, it may compel the owners to plant trees.

In effect, the statute decentralizes substantial powers to local governments through the environment committees. However, the National Environment Management Authority can intervene should local institutions and individuals fail to fulfill their responsibilities. The statute links into the Forests Act by providing for management of all forests in accordance with the principle of sustainable development. It establishes a relationship between the National Environment Management Authority and the lead agency in forestry (currently the Forestry Department). This relationship is now unclear because the new forest law establishes many 'lead agencies'.

Although most of the local government levels have environment committees in place, they remain largely ineffective because members are not compensated. The environment thus remains a popular subject only during public rallies: actually preventing environmentally harmful timber harvesting or cultivation of wetlands is still highly unpopular. Local political leaders find it appropriate to look the other way if their constituents are violating the environmental laws.

Uganda Wildlife Statute, 1996

The statute defines wildlife as 'any wild plant or wild animal of a species native to Uganda' and vests 'ownership of every wild animal and wild plant existing in its wild habitat in Uganda' in the government for the benefit of the people of Uganda. This seems to centralize management of all natural vegetation under the Uganda Wildlife Authority. In practice, the authority manages wild animals (even on private lands), but manages only plants that are in national parks and wildlife reserves.

In essence, the responsibilities of wildlife management are centralized. However, the statute also empowers local government to appoint committees to advise the Uganda Wildlife Authority on the management and utilization of wildlife within the local jurisdiction. Such committees play an advisory role, and getting them to advise a central authority is often more easily said than done.

Lessons learned

Ambiguities in the laws

The ambiguities in the laws have resulted in struggles between the Forestry Department and the local governments to control the forest reserves. Relations between the central government agency and some districts are strained, and cooperation in forest protection remains at arm's length. The Forestry Department is demonized at many political rallies, especially during elections. Local political leaders often lend covert support to unscrupulous individuals in their bid to assume personal ownership of reserves, and there is persistent pressure, often from individuals, to degazette the reserves for other land uses.

The forest sector reform process

The government decided to reform the forestry sector in 1998, although stakeholders had been discussing the subject since 1995. A sector review was carried out, followed by participatory processes, which resulted in the Forest Policy of 2001 and the National Forestry and Tree Planting Act of 2003. These instruments provided for institutional reform, leading to institutionalization of responsibilities for managing forests to four principal actors: the National Forestry Authority, the Uganda Wildlife Authority, the local governments and private forest owners. The responsibilities of local officials and private forest owners have been vested in the district forest services: the forest management arm of the local governments.

The most difficult aspect of this reform process has been reaching agreement on the institutional set-up of the National Forestry Authority and the district forest service. As a result, staff morale within the Forestry Department collapsed, and it became very difficult to maintain discipline in forest management at field level. Opportunists in the private and public sector took advantage of this state of flux to wreak havoc on the forests. Local communities used their voting power to coerce politicians into supporting their illegal encroachment on forest reserves. The forest resource suffered so heavily that many people were asking whether the reform (which is far from complete) had been worth the damage.

Financing sustainable forest management

Public financing

Many forestry activities are being funded by the government and its development partners. This kind of financing is now steadily being transferred to budget support and will be spent through basket and sector-wide mechanisms. Unfortunately, forestry is likely to lose out to education, health, roads, agriculture and other sectors that have higher priority for the government. This state of affairs is likely to be similar at the local levels when decentralization of forest management finally takes root.

Notwithstanding the autonomous decision-making powers of local governments, most of their budget (up to 90 per cent in many districts) still comes from the central government through grants and donor funds. Therefore, their priorities mirror those of the central government, and even districts whose forests are a potential source of revenue are hardly investing in the development of the resource.

In fact, many local officials complain that when the central government created local forest reserves, it did not transfer a corresponding amount of resources (salaries and operational funds) to enable effective management. Since there was little to harvest in these reserves, they remained largely unmanaged, and worse, many suffered further encroachment.

However, donor funding can now be channelled directly through the district governments. Forestry has benefited; but this kind of funding is limited to a few districts with special circumstances, such as refugees and susceptibility to natural disasters. However, when the new District Forestry Service becomes fully operational, there is more chance that more public funding will go to the districts. They will also be able to collect more forestry revenues.

Revenue-sharing with local governments

In 1996 the government allowed the Forestry Department to leave 40 per cent of the gross revenue collected from forest products directly with the districts. It was hoped that this would encourage reinvestment of forestry revenues in the local areas where it had been generated. And if properly collected and reinvested, the share for the district governments could have a substantial impact. For example, 40 per cent of the revenue from royalties on roundwood for timber from pine plantations alone for July–December 2003 was 400 million Ugandan shillings (US$1 was equivalent to 1960 Ugandan shillings in November 2003).

In districts with appreciable revenue from forests, the popularity of the local district forest officer increased; but after complaints from the communities living near forests that they were not seeing the money, district officials started to finance some forestry activities – mainly vehicles for the officers and subsistence allowances for a few staff. Some districts have begun funding nurseries at local levels from development grants that come from the central government. Progress is steady, although the pace could have been faster.

Community investment

Where there were clear financial benefits for individuals, they invested their own resources. In Bushenyi district, the local timber harvesting association tracks down illegal timber dealers and reports them to Forestry Department staff. They are motivated by the increase in timber prices, which means more profits for their members on the sale of legally harvested timber.

The need for public financing remains if people are to see the long-term, less tangible benefits and are to dig into their pockets to invest in forest management. Given their poverty, the argument must be more than just the 'greater good' premise. Income-generating and community projects (such as school furniture or a local clinic) have been tried; but there is no documented evidence that they inspire good forest management.

CONCLUSION

Decentralization as a principle and practice of managing society's business has taken hold in Uganda; but foresters are still apprehensive, and the local governments do not trust the intentions of the progressive transfer of forest reserves as the law provides. The indisputable meeting point of all levels of government will be at the grassroots community. Therefore, activities at this level are likely to gradually remove the mistrust, particularly if the focus is on improving people's lives.

Landownership seems to be closely intertwined with user rights. Most people do not want to have only rights to use the forest; they also want to own the land and to be able to change land use at will. In pursuit of this objective, voters are taking advantage of their powers to coerce politicians into supporting them in their quest for private ownership of the lands upon which forest reserves and other protected areas are located. Local officials are more susceptible to this coercion than their colleagues at the centre. Therefore, the solution may lie in keeping critical forest ecosystems under central control with the National Forestry Authority (compare Chapter 6). However, more forest reserves should be given to local governments to make it worth their while to plan and manage them. The current sharing of 5000ha does not give the 56 district governments incentive to pursue sustainable management.

Responsibility for the local forest reserves was given to local governments in 1998; but corresponding resources for their management did not follow. Technical staff remained at the centre, and the districts were to call on them for advice. But the districts wanted their own staff from whom they could demand services. Although the Local Government Act empowered them to hire staff, forestry was not a top priority in the face of a financial squeeze. The District Forestry Service under the current law will make it possible for the government to provide salaries to the basic forestry staff at the local level.

An incentive scheme is needed for private owners of forests. Estimates show that 50 per cent of the natural forests are degraded. Investment must therefore be directed at restoration work and responsible management of what natural forest remains. Affirmative action in providing forestry management grants to local governments is necessary because forests produce important public benefits such as watershed protection and soil conservation.

Nevertheless, some districts are beginning to finance forestry for community development. They are starting nurseries (albeit very small ones) to supply tree seedlings to local people, often free of charge. If the central government is serious about decentralizing forests, it will be necessary to condition some of its revenue-sharing on developments like these. This is how it started with the other sectors.

The decentralization process in Uganda is succeeding in education, health, roads and agriculture – sectors that are top priority for the central government. Forestry is low on the list for local governments as well. Unless the forests are decentralized, together with the resources to manage them, the local governments are not likely to pay attention to them beyond harvesting them for revenue.

The process of decentralizing forest management should acknowledge that mistakes may be made by governments at both levels. What the decentralization process should do is to limit the extent (temporal and spatial) and concentrate on building confidence among the major players. In due course, the local governments will build their capacity and confidence, and the central government will come to appreciate the benefits of sharing forest management responsibility.

In any case, forests in fragile ecosystems often do not yield sufficient profits to compete favourably with land-use sectors such as agriculture. Therefore, the central government will have to target funding, whether through local governments or through semi-autonomous bodies such as the National Forestry Authority or the Uganda Wildlife Authority. That is one of the reasons why the authority to hold forests in trust for the people of Uganda was placed at the two levels of government.

The ongoing process of revising the constitution may remove the current ambiguities in the laws governing forests. This will then make it more plausible for the responsible institutions to work diligently at capacity development for sustainable forest management.

References

Kato, D. (1997) 'Uganda's experience in the use of service delivery surveys'. Paper delivered to Eighth International Anti-Corruption Conference, Lima, Peru

Ministry of Water, Lands and Environment (2001) Uganda Forestry Policy. Ministry of Water, Lands and Environment, Kampala

Chapter 11

Decentralization of Federal Forestry Systems in Ghana

Oppon Sasu

INTRODUCTION

Decentralization of governance and forest management in Ghana was initiated prior to 1878 during the colonial era to empower traditional authorities (chiefs), primarily to solicit their support in enforcing colonial policies. However, after independence in 1957, successive governments increasingly strengthened state control over local governance and natural resources. Currently, the focus of the decentralized system is the promotion of consultative mechanisms, devolution of power, competence and resources to the district assemblies. Although decision-making, planning and execution of development projects at the district level are the prerogative of district assemblies, the decentralization process has suffered from the central government's reluctance to completely decentralize important revenue-generating sectors, such as forestry and mining. In addition, the participation of civil society, non-governmental organizations (NGOs) and community-based organizations (CBOs) in the process has been weak. Empowerment and capacity building of local communities to effectively manage their natural resources must be accorded the necessary priority by government.

LANDOWNERSHIP IN GHANA

In Ghana, landownership, land rights and tenures are administered in a plural legal environment, with customary laws and norms operating alongside statutes. There are several kinds of customary landowners in the country:

- stools: stool land is vested on behalf of the community represented by the chief, and members of the landholding group have usufruct rights, equivalent to a freehold;
- clans: when families merged to form clans, the individual family or household lands came under the authority of the newly formed clan,

which later merged to form larger tribal groups or federations under one traditional authority;

- families;
- *tindanas*: the spiritual or religious heads of the land in northern Ghana; and
- *tendamba*: the owners of the land and forest groves in some parts of northern Ghana, usually, the first settlers.

Together, these landowners own about 78 per cent of the total land area in Ghana. Of the remainder, the state owns 20 per cent, and 2 per cent is held in dual ownership – the legal interest in the government and the beneficiary interest in the community (Kasanga, 2002). The rights of ownership of all customary lands, including forest reserves, reside in the original landowners; however, management rights of commercial natural resources such as timber, precious minerals and salt are the responsibility of state agencies and are governed by legislation. The uniqueness of traditional landownership in Ghana has had an enormous influence on the evolution of decentralization in the country.

Widespread customary landownership means that the role of local users in resources management is recognized. Nonetheless, not much has been done to promote collaborative resources management because of the state's reluctance to devolve management authority to the local communities. Since the early 20th century, succeeding governments, including the colonial administration, have attempted to shift control or supervision over disposition of lands from traditional owners to the state, largely because of the perceived misuse of land by traditional authorities and the general decline of state authority in the past. Legislation has therefore granted sweeping state control over natural resources (including timber and precious minerals), land use, enhanced powers of expropriation and the assumption of the managerial and fiduciary powers of the stools with respect to unencumbered land. It was assumed that local people had no knowledge of resources management and no interest in its conservation or protection (Kotey et al, 1998).

GHANA'S DECENTRALIZATION POLICY AND IMPLEMENTATION PLAN

Historical perspective

Local government was introduced in Ghana in 1878, during the colonial era, by the British. The purpose of this decentralization exercise was to provide a legal basis for traditional authorities to carry out limited local government functions, including judicial and legislative activities and resources management. These functions were modified during the 1940s to include the appointment of educational and technical specialists for administrative functions and to ensure a greater community voice by making chiefs the presidents of the local government councils (GoG, 2002). Thus, decentralization focused on empowering traditional authorities to ensure that they supported colonial policies.

However, after independence in 1957, state control over local governance and natural resources increased. In fact, by 1965, Ghana had a very strong central administration in the office of the president, with 182 weak urban and local councils and 161 ineffective district councils. The state passed a series of laws entrenching centralized management of stool lands, the granting of timber rights and the collection and disbursement of revenue on behalf of landowners. This situation led to conflicts between the state and landowners over benefit-sharing and utilization of natural resources. Because of this increasing centralization of land and natural resources administration, local communities lacked the capacity to manage their forestland and natural resources. In addition, the state passed laws to reinforce its control over natural resources to the exclusion of other stakeholders, particularly the chiefs and local communities, because it needed revenue to meet the increasing level of government expenditures.

By the early 1970s, the rapidly increasing population and expanding local government activities made it virtually impossible to run the local government system from the office of the president, and therefore a four-tier structure consisting of regional, district and local councils and village development committees was established. The district councils were made the main focus of local government. The 1970s and the early 1980s were a period of political instability, with three coups and several attempts to overthrow the government. The combination of this political volatility and a poor economic climate thwarted the implementation of effective decentralization and economic development.

By the mid 1980s, the public forestry agencies' ability to manage natural resources had deteriorated. Each year during the 1970s and early 1980s, an estimated 20,000 hectares (ha) of valuable timberland were lost to wildfires, agriculture and logging (FAO, 1988). In 1982, Ghana began to engineer a political and economic recovery. Under an economic recovery programme supported by the International Monetary Fund (IMF), reforms were implemented to improve the Ghanaian economy by addressing inflation, external deficit and declining output. The development strategy of the country changed from import oriented to export oriented, with increasing private-sector participation and reduced government involvement in direct production through the divestiture of state-owned enterprises.

Despite the success of this and other programmes in reversing the economic decline of the 1970s and achieving a positive rate of economic growth, income per head rose by less than 2 per cent per annum because the population continued to grow (World Bank, 1993). During the period, about 36 per cent of Ghana's population was estimated to live in poverty.

In order to alleviate poverty and increase the voice of local communities in governance, the government initiated a comprehensive review of the local government structures with the enactment of the Local Government Law of 1988 (PNDCL 207). The focus was to promote consultative mechanisms and the devolution of power, competence and resources to the district level, and to ensure that all ministries, departments and agencies of government decentralized their operations and activities to the district level (GoG, 2002).

The decentralization process was further strengthened in 2003, when the government put in place the three-year National Decentralization Plan; with the participation of a wide range of stakeholders, it seeks to harmonize capacity-building activities, coordinate the decentralization programme and develop effective strategies for decentralization. According to the Ministry of Local Government and Rural Development (2002), the outstanding issues include the following:

- resolving the outstanding difficulties created by the establishment of public services for forestry, health and education as major functions are decentralized;
- enacting legislation to clarify the function and related powers to be exercised by ministries, departments and agencies at various levels; and
- developing the capacity of sector ministries to transfer power and functions.

Decentralized institutions, governance and forest management

Local institutions and forest management

One feature of the decentralization process in Ghana is the continued resistance of government agencies. In addition, central government has also hesitated to enforce the decentralized system, especially for revenue-generating sectors such as forestry and mining (Amanor and Brown, 2003). It is not yet clear whether decentralization will *de facto* shift control over allocation and administration of forest resources to the district assemblies or whether the assemblies will ensure that the forestry sector's programmes are reflected in the development aspirations of the district.

Currently, the district assemblies, which consist of 67 per cent elected representatives and 33 per cent government appointees, are responsible for the overall development of the districts and are more than an interested party in all development activities, including the management of natural resources. Theoretically, all line agencies of the central government, such as the Forestry Commission, within each district are expected to report to the assembly. However, personnel and logistics problems still hinder implementation in many districts.

Almost all of the district assemblies have established subcommittees on the environment to deal with natural resources management and other environmental issues; relevant agencies, including the Forestry Commission, are represented on these subcommittees. Many assemblies, through their environmental subcommittees, have passed environmental laws, such as restricting chainsaw operations and charcoal burning.

At the community level, unit committees form the base structure of the new local government system. A unit is normally a settlement or group of settlements with a population of 500 to 1000 in rural areas and 1500 for urban areas. Unit committees, being in close touch with the people, play the important roles of education, organization of communal labour, revenue-raising, environmental protection and implementation of self-help projects. The Forestry Commission has formed community forestry committees at the unit level to enhance community empowerment and participation in resources management.

The weaknesses of local institutions and their influence on collaborative forest management approaches are a shortfall of the decentralization process. For example, Amanor and Brown (2003), in a review of the impact of decentralization and local environmental management, observed that the performance of the decentralized structures has been weak probably because of chronic under-resourcing and its resulting incapacity.

Tenurial systems and forest management

The impacts of customary tenurial systems and the performance of indigenous land institutions on sustainable resource utilization and management are enormous. For example, it has been noted that indigenous institutional arrangements ensure better accountability to local communities and villagers than the public-sector agencies and state machinery. This is one of the reasons why the Ghana poverty reduction strategy recommends the promotion of customary systems as a mechanism for sustained poverty alleviation. In addition, there are more effective checks and balances at the community level (see contrasting views in Chapters 4 and 10).

Generally, access to land and security of tenure are the prerequisites for any successful programme to reduce poverty in the country: insecure tenure is endemic in most regions and impinges upon both poverty reduction and economic growth. The failure to provide for the protection of land rights, domestic use rights of forests and prevention of abuse of traditional and institutional procedures places the vulnerable groups in society, including the poor, the illiterate, women, migrants and tenants, at most risk.

One major issue that the decentralization process has not adequately addressed is tenurial conflict and its effect on resources management at the forest management unit and local community levels. Farmers are the main custodians of forest resources outside reserves; yet management and utilization rights are vested in the state, resulting in the extremely high rates of illegal logging and chainsaw lumbering in the country. For example, a recent study on the performance of the wood industry estimated that the total harvest for 1999 was about 3.7 million cubic metres, which is almost four times the annual allowable cut. Of the total amount of timber harvested, about 0.9 million cubic metres (24 per cent of total harvest) was illegally harvested by timber utilization contract holders, and another 1.7 million cubic metres (46 per cent of total harvest) was harvested by illegal small-scale chainsaw operators (Birikorang, 2002). The high rates of illegal logging and chainsaw operations arise mainly from the lack of equitable benefit-sharing and the lack of active involvement of farmers and local communities in the management of forest resources.

One incentive for the illegal chainsaw operations to the local communities is the prompt payment of benefits irrespective of the amount of money involved. The chainsaw business thus comes closer to community households, at least in decision-making on exploitation of trees and meeting the immediate and short-term needs of households for timber. The government is currently undertaking a consultation process with the chainsaw operators and local communities as a long-term solution to the problem. Steps are also being taken to ensure that resource-owning chiefs, fringe communities and rural communities effectively collaborate with the Forestry Commission in the protection of the forests and also to ensure that these stakeholders receive adequate and remunerative benefits for their efforts. However, the full participation of civil society in forest management has been hampered because of under-pricing of land and forest resources by the state agencies, which makes it less attractive for civil society to participate in resource protection and unprofitable for government to embark on effective public awareness and education programmes.

The role of traditional authorities in local communities and forest management

During the past, traditional land and tree tenure systems were incorporated through cultural norms, traditions and beliefs into everyday life. This ensured that forests and trees were protected. However, increasing population pressure, land scarcity and cultural changes have reduced the effectiveness of these systems. According to Danso (2000), there was no evidence of chiefs' holding customary rights over on-farm timber before the 1920s. However, development of the timber trade led to the system in which all rights to land assets that are regarded as natural were vested in the chiefs through the Concessions Act of 1939 and the Local Government Act of 1951. Amanor (1996) reiterates that the customary rights of chiefs in timber trees on farms are an invented tradition that gradually emerged from the colonial period between the 1920s and 1940s as a result of an alliance between the chiefs and the colonial government. The rationale behind strengthening the traditional system was that the traditional authorities were thought to control land for the benefit of the subjects or members of the stool, clan, company or community (Acquaye and Murphy, 1973).

Even though land ownership has changed from traditional to state control in many countries, traditional ownership of land in Ghana has remained unchanged over the years. The continued ownership of land by traditional authorities has, however, not resulted in increased control over forest resources largely because succeeding governments in the country have strengthened state control over resources while maintaining traditional ownership of land.

For example, during the creation of forest reserves in the country under the Forest Ordinance during the early 20th century, landowners were free to decide whether they would manage the forest reserves by themselves (under the supervision by the forestry agencies) or have it managed on their behalf. Two types of forest reserves were thus in existence: the so-called ordinance reserves (managed directly by government) and by-law reserves (created by chiefs' by-laws and managed by local chiefs under the direction of the forestry agencies). It was significant that, by 1939, more that 80 per cent of all the reserves created in the country were established by native by-laws enacted by the chiefs. Even though these forests were to be managed by local communities under state direction, the reserves were, instead, managed by the state with little or no input from the communities. Thus, the management of by-law reserves was compromised by encroachers and illegal loggers.

It was generally perceived that forest reserve management under by-laws meant that if the local administration had the necessary resources and commitment, there would have been no need for government intervention. Unfortunately, numerous spurious activities were carried out in many reserves created under by-laws. This led to the eventual placement of all reserves under ordinance between the 1940s and 1950s without consultation with the local communities who owned the land and forest. Thus, as stated earlier, the major shifts in forest management have been from initial attempts at colonizing through collaboration to 'dictatorial' rather than 'democratic' management.

A World Bank consultancy team reviewing the forestry sector performance in 2001 remarked:

> *While Ghana has been among the first countries in Africa to recognize the role of local people in resources management, the debate on community participation in forest management has increasingly shifted towards rhetoric in the past several years. Current work by specialized forest institutions seems to focus on sharpening instruments for implementation of current regulations, overlooking the fact that existing regulations do not adequately reflect stated participatory management policies and benefit-sharing arrangements. Even social responsibility agreements (SRA), the most advanced and specific framework instrument for communities to negotiate with logging companies their in-kind contribution of local infrastructure, leave communities with very little control and negotiating power (Ryan, 2001).*

Prior to 1994, the role of traditional authorities and local communities in forest management had declined and was limited to consultation in resource allocation and a small share in benefits from the forest resource. Even though the stool, the traditional council and district assemblies are named as the recipients of the royalties from timber harvesting in the 1992 constitution, disbursement of such revenue is often delayed. In addition, the royalty rates are low, and illegal activities reduce payable royalties. None of the recipients are, by law, obliged to spend any of the monies received from royalties on the communities where the resources came from. For this reason most community members complain of the lack of benefit flow from the forest resources. This situation led to distrust of the Forestry Commission, apathy and, in a few instances, connivance of some chiefs and local communities with illegal operators.

A new forest and wildlife policy promulgated in 1994 indicated a shift in government policy, from authoritarian control to stakeholder involvement. This new approach is marked by some 'guiding principles':

- enhancing the rights of people to have access to natural resources for maintaining a basic standard of living and their concomitant responsibility to ensure the sustainable use of such resources;
- incorporating traditional methods in resources management in national strategies where appropriate;
- promoting the importance of appropriate and efficient land use and security and land tenure for sustainable development; and
- identifying the need to develop a decentralized participatory democracy by involving local people in matters affecting their welfare.

Implementation of forestry decentralization

Implications of the programme for forest management

The Ministry of Lands and Forestry has recently shifted from a government-led system to a civil society–government collaborative management approach. It recognized the importance of involving civil society in land and forest management because of the uniqueness of the country's tenurial systems and the strong interest and rights of civil society and other stakeholders in forest resources management. According to Amanor and Brown (2003), resources will be more efficiently, equitably and sustainably managed if decision-making is brought closer to the primary users.

The weakness of the forestry decentralization process was that it was driven by the Forestry Commission, which set its own benchmarks for decentralization. Until the late 1990s, local communities and other stakeholders were barely involved in forest management. Although policies on forest decentralization are well developed in Ghana and some functions and decision-making have devolved to the district forestry offices, the involvement of civil society, NGOs and CBOs has lagged.

In several instances, decentralization of decision-making has not resulted in efficient resource utilization because it was not accompanied by empowerment and capacity-building of the communities to manage the resources effectively. For example, in order to control illegal chainsaw and logging activities, all chainsaw operators were required to register with their district assemblies. This action did not produce the desired effect and was revoked because some district assemblies, traditional authorities and local forestry officials abused the system, using it as a revenue-generating venture or as an avenue for rent-seeking.

The government's accelerated decentralization has not yet resulted in a positive effect on local people's attitudes to forests and trees, or built trust in public forestry agencies. An effective strategy is needed. In particular, policy and legislative constraints associated with the distribution and/or sharing of forest revenue, including resource ownership, resource allocation and benefit-sharing among stakeholders, should be addressed. If the issue of equitable benefit-sharing is left unresolved, it may actually increase poverty among forest fringe communities for the following reasons:

- distribution of resource rent is skewed in favour of the timber industry and governmental institutions to the disadvantage of resource landowners and forest fringe communities;
- within communities, stumpage distribution does not trickle down to households;
- landowners and tenant farmers have no incentive not to collaborate with illegal timber and chainsaw operators; and
- landowners and farmers have no property rights over trees or cannot protect their rights.

Making forest management more responsive to local communities

The Forestry Commission's programme to address questions of rights, access to services and benefit-sharing builds upon earlier and ongoing work, such as the implementation of the service charter and collaborative forest and wildlife management programmes. The commission has also facilitated the establishment of a network of forest fora, with the support of the UK Department for International Development (DFID) and the United Nations Food and Agriculture Organization (FAO), which envisages the creation of fora at district and national levels to strengthen civil society in support of sustainable forest management. The fora were also initiated to promote changes that would benefit the poor.

Since 2000, the government has put in place a strategy to address past lapses in the forest decentralization programme because, from a socio-economic perspective, forest management has not brought substantial benefits to the people at the community level. People have not been given the opportunity to participate in the decision-making process, nor are their domestic needs and aspirations taken into consideration in the formulation of policy and management schemes for their forests. To address this shortfall, the Forestry Commission, with DFID and World Bank funding, is promoting four programmes:

1. Governance and institutional development: measures to improve transparency and accountability within public sector agencies have been initiated. Strategies are being developed to effectively implement the measures.
2. Decentralization and capacity-building: at the grassroots, local and district assembly levels, existing structures are being strengthened to foster local development, build sustainable partnerships among all stakeholder groups and give them a voice.
3. Local community involvement: local community participation in forest management is being more rigorously enforced.
4. Poverty reduction: especially at the rural community level, through enhanced revenue generation, wealth creation and improvements in people's livelihoods and income base are being strengthened.

The operational plan to achieve those goals is the promotion of community empowerment and strengthening of civil society participation in forest management and decision-making. It means more than just providing infrastructure and skills; it involves strengthening organizations' abilities to do their work in collaboration with, or in opposition to, other actors and forces. It also involves paving the way for group formation and collective action.

Under the operational plan, local communities will receive compensation for their work, as well as equitable distribution of the forest rent from trees harvested. According to Smith (1999), the forest decentralization programme is aimed at re-establishing local communities as primary beneficiaries of forest management with an absolute right to benefit from the wise stewardship of their resource. The current decentralization programme is also expected to improve the cost-effectiveness of forest management prescriptions through consultation, needs assessment, investigation, synthesis and consensus-building.

In addition to the above, the government is looking at ways to involve local communities and the civil society more closely in forest management activities:

- Local communities and district assemblies will be involved in resource protection and felling inspections for off-reserve areas. NGOs will also be involved in policy formulation and monitoring resource exploitation and management.
- Benefits from resource rent and other revenue will be disbursed directly to local communities and other primary stakeholders who are not currently benefiting directly from the existing benefit-sharing framework.
- Efforts will be made to ensure that the 55 per cent of the revenue share to district assemblies is used for projects within the communities where

the resources were exploited. This will ensure greater cooperation of the local communities in resource protection.

TRANSITIONAL ISSUES RESULTING FROM DECENTRALIZATION

Ghana is one of the few countries in the west-central African sub-region undertaking a comprehensive decentralized system of governance. A district assembly common fund has been set up in which government disburses not less than 7 per cent of the total revenue. The district assemblies are the agents of development, using the common fund and other revenue generated at the district level.

Decentralization has had enormous impacts on infrastructural development at the local level. Both the government's and the forestry sector's decentralization programmes have also had significant impacts on controlling wildfire and preventing deforestation and illegalities, largely because of greater participation of local communities in such programmes.

However, decentralization has failed to make the desired impact, mainly because of the high demand for wood and the low penalties for illegalities. Again, poor capacity at the district forest operations level has prevented detection of illegal activities and allowed connivance of forestry field officers with illegal operators, thus exacerbating the situation.

Another factor that has reduced the impact of decentralization is the low level of interaction among stakeholders. This led not only to widely differing views of how decentralization should and could work, but also to a considerable gap between the rhetoric and substance of policies promoting community involvement in forest and wildlife management. For example, the transfer of responsibility for off-reserve timber resources management from district assemblies to the Forestry Department (now the Forest Services Division of the Forestry Commission) in 1996 – an apparent reversal of the decentralization process – tended to strengthen the concession system, which favoured large-scale wood processors but denied small-scale producers access to forest resources. The apparent increase in 'illegal' timber felling reflects the criminalization of these small-scale wood operations.

A central lesson of both the national and forest decentralization processes is that the revenue-generating sectors have still not been decentralized. For example, within the Forestry Commission, regional and district forestry officers cannot allocate or sell even a single timber tree to individuals, communities or firms without the approval of the chief executive of the Forestry Commission. Nor can they decide on the disbursement of revenue collected from forest products. Losers are the forest-owning communities, who cannot ensure an equitable flow of benefits to themselves. Thus, the financial and budgetary outcomes from decentralization, particularly in terms of forest revenues, costs of forest administration, taxes and penalties collected, does not seem to have gone far enough. Both government and the Forestry Commission are apprehensive about relinquishing financial and administrative control – a major reason for the lack of sustainable impacts.

The decentralization programme has sought to alter incentives in the sector, so that forestry professionals are encouraged to adopt a service role and communities and district assemblies are rewarded for pursuing sustainable forest management and conservation. However, the logging companies have begun complaining that landowners and forest-fringe communities have become more aggressive in ensuring that their rights are met, and in some cases made excessive demands, under the social responsibility agreement, which have in some instances stalled timber operations.

The empowerment of local communities by the decentralization programme has resulted in numerous cases of conflicts:

- increased discord between stakeholders and a lack of dialogue and communication;
- marginalization of tenant and settler farmers and other disadvantaged groups in resource allocation – this is a major problem, even when it is only a perception;
- inequitable distribution of benefits to local communities since the few benefits from resource utilization do not trickle down to resource-owning communities, but end up with the chiefs;
- lack of recognition of property rights; and
- weaknesses of state and traditional resources management institutions, which results in poor implementation and monitoring of programmes.

The decentralization process has empowered the district assemblies to set up arbitration committees to resolve conflicts. For forest offenses, the manual of the Forestry Commission states that if, in the opinion of the district forest manager, a forest operation is not being carried out in accordance with regulations, the officer may suspend the operation until corrective measures are complete. However, the law courts are the final arbiter in all cases.

Lessons learned

One of the major failures of the decentralization process is that the 1992 constitution requires forest royalties to be shared in the following proportions after a 10 per cent administrative charge has been levied by the administrator of stool lands:

- district assemblies: 55 per cent;
- stool chief (or alienation holder): 25 per cent; and
- traditional council: 20 per cent.

The beneficiaries, especially the district assemblies, have few responsibilities for forest management but derive huge benefits from the forest revenue. None of these recipients are, by law, obliged to spend the monies on the communities where the resources came from. Thus, benefits do not accrue to local populations, who are commonly alienated from management responsibility. Farmers and forest fringe communities are responsible for controlling wildfire and preventing deforestation and illegalities, yet do not receive significant benefits, resulting in their increasing poverty. Thus, for the decentralization process to be more effective there must be a comprehensive legal and policy review to promote the rights of farmers and other marginalized groups and to enhance their institutional role in forest management.

Some examples of successful decentralization efforts have occurred in districts with ecotourism potential. Detailed agreements were signed between the Forestry Commission, district agencies and traditional rulers regarding the roles and responsibilities of stakeholders and arrangements for equitable benefit-sharing from the revenue generated from natural resources and associated income. An example of this is the Agumatsa Wildlife Sanctuary benefit-sharing arrangement, in which 57 per cent of revenue goes to the Wli (forest-fringe) community, 23 per cent goes to the wildlife division and 20 per cent to the Hohoe District Assembly (Agyeman et al, 2003). However, these types of arrangements are individual cases and are not backed by the necessary legislative framework. The lesson is that for the decentralization process to succeed, it should be community driven, flexible and transparent. Conscious efforts should be made to empower communities and guarantee their rights to equitable benefit-sharing.

Another lesson is that local community and civil society capacity needs to be built. Despite their perceived weaknesses, customary tenurial systems and indigenous institutions are better able to ensure accountability to local communities and villagers than the state resources management machinery, which is still largely centralized when it comes to administrative and financial management of resources. Strengthening local community groups and other civil society groups for effective resources management is more than just providing infrastructure and skills. These organizations must be able to do their work in collaboration with or in opposition to other actors and forces, as well as pave the way for group formation and collective action (Rhein, 2002). Many civil society groups with an interest and responsibility for forest management issues are springing up, and government must promote their activities.

The commitment of development partners influences the degree of success. The decentralization process has achieved its greatest gains where development partners have allowed the process to be country led, even if the programme proceeds slowly. According to Tsikata (1999), development assistance can lead to growth only if a good policy framework supports it. In the particular example of Ghana, development assistance to the decentralization process has resulted in the desired changes because the government was willing to make difficult decisions and implement tough economic, structural and institutional reforms.

Currently, the roles and responsibilities of institutions and organizations supporting the decentralization process are evolving. The premise for the development of regional and local structures of government is that effective institutional and organizational development is about good governance, control and communication, and performance. It is apparent that good governance, especially transparency and equitable sharing of resources and benefits, is critical to the well-being of local communities and seems to be a catalyst for poverty reduction. The governance challenge is not just a technical problem – that is, a matter of systems; it has strong political dimensions. A prerequisite to good governance is the clarification of roles and responsibilities between actors in the sector.

Conclusion

In order to be more efficient, the forestry sector in Ghana needs to adopt a pro-poor agenda in conformity with the decentralization objective of the country. Currently, the Forestry Commission has developed a 'pro-poor growth agenda' strategy to support its own decentralization programme and to ensure that local communities receive equitable benefits from forest resources management. The pro-poor growth agenda strategy of the forestry sector has several objectives:

- to generate a broader range of perspectives as inputs into the policy process;
- to enhance the likelihood of effective implementation of sector reforms by broadening the range of actors involved in implementation;
- to increase the political viability of sector reforms through broadened participation and commitment; and
- as a result of the above, to increase the likelihood that sector reforms will have a significant social and economic impact and reduce poverty.

The pro-poor agenda of the forestry sector draws heavily from the Ghana Poverty Reduction Strategy (GPRS) programme. The goal of the programme is to achieve equitable economic growth and accelerated poverty reduction within a sustained democracy. Emphasis is on stabilizing the economy; laying a sustainable foundation for accelerated and equitable economic growth; and reducing geographical gaps in poverty, including access to social services and ameliorating conditions of extreme poverty. The following policy thrusts are being pursued to achieve these goals:

- ensuring macro-economic stability for accelerated growth;
- increasing production and gainful employment opportunities;
- facilitating direct support for human resource development and provision of basic services;
- expanding special programmes to support vulnerable groups; and
- enhancing good governance.

A strong civil society and an active private sector are central to successful sector reforms and sustainable development. The approach to strengthening civil society should be adapted to local contexts.

Finally, forestry offers real opportunities for sustainable economic growth, poverty reduction, improved governance, enhanced public-sector performance and the development of empowered communities and a competitive forest industry. To utilize its potential for national development, the forestry sector needs to move from rhetoric to an effective programme that aligns the implementation of its pro-poor growth agenda with the government's poverty reduction strategy.

References

Acquaye, E. and Murphy, M. C. (1973) *Land Use, Land Tenure and Agricultural Development in Ghana (with Two Case Studies, Offinso and Kumbungu Areas); Contemporary Changes in Agrarian Structure*. UN/FAO Studies, Land Administration Research Centre, Department of Land Economy, UST, Kumasi, Ghana

Agyeman, V. K., Kasanga, K. R., Danso, E., Marfo, K. A., Whiteman, A., Asare, A. B., Yeboah, O. M. and Agyeman, F. (2003) *Equitable Forest Reserve Plantation Revenue Sharing in Ghana*. Report for UN Food and Agriculture Organization (FAO), Rome

Amanor, K. S. (1996) *Managing Trees in the Farming System: The Perspective of Farmers*. Forest Farming Series No 1, Forestry Department, Accra, Ghana

Amanor, K. and Brown, D. (2003) *Making Environmental Management More Responsive to Local Needs: Decentralisation and Evidence-Based Policy in Ghana*. ODI Forestry Briefing No 3, April 2003, Overseas Development Institute, London

Birikorang, G. (2002) *Ghana Wood Industry and Log Export Ban Study Report 2001*. DfID and Forestry Commission Ghana, Assembly Press, Accra, Ghana

Danso, E. Y. (2000) *A Review of ODA Support to Participatory Forest Management in the High Forest Zone of Ghana: The Collaborative Forest Management Programme of the Ghana Forestry Department*. Case study report prepared for the Review of Overseas Development Administration (ODA) Support to Participatory Forest Management in West Africa, IIED, London

FAO (United Nations Food and Agriculture Organization) (1988) *An Interim Report on the State of Forest Resources in the Developing Countries*. FAO, Rome

GoG (Government of Ghana) (2002) 'Economic policy framework'. Paper prepared by the Mini-CG Secretariat for the Eleventh Meeting of the Consultative Group (CG) for Ghana, Accra, April 8–10

Kasanga, K. R. (2002) *Integrating Land Issues into Poverty Reduction Strategies and the Broader Development Agenda: The Case of Ghana*. Regional Workshop on Land Issues in Africa and the Middle East, Kampala, Uganda

Kotey, E. N. A., François, J., Owusu, J. G. K., Yeboah, R., Amanor, K. S. and Antwi, L. (1998) *Falling into Place: Policy that Works for Forests and People*. Series No 4, International Institute for Environment and Development, London

Ministry of Local Government and Rural Development (2002) *Decentralisation in Ghana: Implementation Status and Proposed Future Directions*. Briefing sheet, March, Ministry of Local Government and Rural Development, Accra, Ghana

Rhein, M. (2002) *Supporting Pro-poor Change in the Forest and Land Sector through Civil Society Strengthening. FSDP2 Concept Note on Civil Society Strengthening, Forestry Commission*, Forestry Commission Working Report, Accra, Ghana

Smith, E. K. (1999) *Developments and Setbacks in Forest Conservation: The New Political Economy of Forest Resource Use in Southern Ghana*. Ministry of Lands and Forestry technical paper, Accra, Ghana

Tsikata, Y. M. (1999) *Aid and Reform in Ghana*. Economic and Social Research Foundation (ESRF), Dar es Salaam, Tanzania

Ryan, P. A. (2001) *Notes on High Forest Management*. Mid-term review report of the Ghana Natural Resources Management Programme (NRMP), Working Report, Accra, Ghana

World Bank (1993) *World Bank Development Report*. World Bank, Washington, DC

Chapter 12

The Push-Me, Pull-You of Forest Devolution in Scotland

Bill Ritchie and Mandy Haggith

Introduction

The 'push down' of power through devolution and decentralization and the 'pull down' of power through action on the ground are reaching an exciting stage in Scotland. This is happening within a political context of devolution in the UK, with the establishment of the Scottish Parliament in 1999 and Scotland's forest estate now in the hands of the Scottish Executive (the government in Scotland). This chapter describes the political 'push down' and the people's 'pull down' of power and tells the story of the people's movement to get access to land and natural resources. The movement has led to the Land Reform Act, which gives communities the right to acquire ownership of land, including woodlands, and to the establishment of more than 100 community woodland organizations. It has also strongly influenced cultural and organizational changes within the national forest and rural development institutions. Local and sometimes radical efforts to pull down power can shape the decentralization process at the national level. Six examples illustrate the range of options for communities seeking to manage local woodlands.

Background, context and rationale

The people of Scotland have brought about a sea change in forest governance over the past few decades, shifting power towards the local communities for whom forests are part of their home environment. It demonstrates how revolutions at the local grassroots level can make a real difference in the way decentralization is implemented by a national government. Grassroots efforts create new possibilities, proving that 'impossible' things can really be achieved, thereby feeding the collective imagination of the country, building confidence in particular new forms of change and generating political will to create conditions that enable such changes to be replicated.

Once, Scotland was almost entirely covered by forest in which bears and wolves roamed; but over several thousands of years this forest was destroyed, degraded and converted to agricultural and pasture lands. By the end of World War I, deforestation was all but complete, and efforts began to reverse the process with a massive state-run industrial afforestation programme. Today the emphasis has shifted towards a vision that includes environmental and social benefits as well as economic ones, and towards a model of reforestation and forest management that involves, rather than alienates, local communities.

Political context

Scotland is a nation of 5 million people that occupies the northern part of the island of Great Britain. Since the Labour party took control of government after a landslide victory in 1997, the UK is undertaking a major devolution process and thereby fulfilling a longstanding political promise.

In 1997, the Scottish people voted for devolution in a referendum, and in 1999, after nearly three centuries of centralized rule from London, the Scottish Parliament was established in Edinburgh. Domestic issues (such as health, education, justice, transport, local government, agriculture, environment and rural development, including forestry) were devolved to the new government in Scotland: the Scottish Executive. Ownership of Scotland's national forests passed to Scottish ministers.

Until recently, despite devolved forest ownership, the national forest estate was still managed by the centralized Forest Enterprise as agent for the UK Forestry Commission. Following a review of the post-devolution experience, the management of the national forest estate was decentralized in 2003 with the creation of the Forestry Commission Scotland, answerable directly to Scottish ministers.

Brief forest history

Original (postglacial) forest cover in Great Britain about 7000 years ago is estimated to have been around 80 per cent. By 1000 years ago, land cover is estimated to have been reduced to 20 per cent. Recent estimates put the total figure of seminatural woodlands at around 2.5 per cent of land cover. Most of the ancient seminatural woodland areas are very important for biodiversity, and some are unique because of their relative isolation from mainland Europe and the strong oceanic influences of the North Atlantic.

The Forestry Commission was established in 1919 to develop a national timber reserve and, in particular, to grow enough timber to supply coal mines with pit props, without relying entirely on imports. Since then, afforestation has steadily increased tree cover to its current level of 16.4 per cent of Scotland. Total forest area (state and private) is now approximately 1.5 million hectares (ha) in Scotland. As a result of the timber orientation of the afforestation programme, the forests are mostly industrial plantations of non-native species. Now, however, the Forestry Commission has a strong commitment to make further increases in forest cover with native woodland.

To achieve its afforestation goals, the commission purchased large tracts of rural uplands and terminated hundreds of agricultural tenancies. Despite continued expansion of the national forest, estate employment in the forest sector dropped considerably over the years, largely as a result of the mechanization of forest operations, expensive labour costs and the availability of cheap imports. Rural communities who depended upon employment in the forest sector were affected, resulting in many 'ghost' villages.

During the 1980s, UK citizens were increasingly concerned about the equity of the government's incentive schemes to private landowners, which included grants and tax breaks, as well as about the impact of the subsequent increased planting on areas of high natural heritage value, especially the peatlands. Between 1979 and 1997, some 100,000ha of the UK's public forest were sold to the private sector. There was substantial public and private investment in the paper pulp industry. Planning and decisions on forestry were made by people living far away from the forests and by landowners from a different social and economic class than the rural population most affected by the decisions. It can be argued that the forest landscape we see today has been imposed upon rural communities rather than developed from within them.

In Scotland, only around 2 per cent of the land is covered by native woodland, and all of the land has been heavily modified by centuries of (mis)management. Scotland also has one of the most skewed and inequitable landownership patterns in the world, with nearly 60 per cent of the land controlled by as few as 1250 owners, many of them foreign nationals and institutions, and 13 per cent owned by 110 members of the British aristocracy (Wightman, 1996, 1999). The Scottish national forests cover 667,000ha (8.5 per cent of Scotland's land area), forming Scotland's biggest landholding and representing 36 per cent of Scotland's woodland and forests. The remaining 64 per cent is privately owned.

The national forest programme

The global discourse on forest governance – Agenda 21, the United Nations Convention on Biological Diversity (CBD), the United Nations Framework Convention on Climate Change (UNFCCC), the Intergovernmental Panel on Forests (IPF) and Intergovernmental Forum on Forests (IFF) – has a strong influence on UK policies. In 1994, the UK responded to its international commitments by publishing *Sustainable Forestry: The UK Programme* (The UK Government, 1994), which brought together various elements from government policies and programmes and set them in the context of international principles and guidelines. The programme covers all forests, including exotic plantations. However, there is now an increasing emphasis on the conservation, management and expansion of seminatural native woodlands. A series of policy statements and guidelines have been issued since 1985 to conserve and extend broadleaf woodlands and native pine-wood. These have been accompanied by targeted financial incentives for the creation of new native woods and for the removal of exotic species from existing native woodlands.

The *Scottish Forestry Strategy* (Forestry Commission, 2000) is the Scottish Executive's forestry policy. It sets out five strategic directions for Scottish forestry:

1. Maximize the value to the Scottish economy of its wood resource available for harvesting over the next 20 years.
2. Create a diverse forest resource of high quality that will contribute to the economic needs of Scotland throughout the 21st century and beyond.
3. Ensure that forestry in Scotland makes a positive contribution to the environment.
4. Create opportunities for more people to enjoy trees, woods and forests in Scotland.
5. Help communities to benefit from woods and forests.

Underpinning these strategic directions are five principles: sustainability; integration; ensuring positive value; securing community support; and recognizing local diversity and distinctiveness. The overall strategy recognizes that the environmental and social benefits sought from forestry do not necessarily generate cash income for the owner. Since Scotland's national forests are managed in the public interest, the Forestry Commission is expected to deliver such social and environmental benefits. The Scottish Executive has allocated an average of UK£66.2 million per year for 2003–2006 towards the cost of running the national forests. Owners of private forests have access to grants to help meet the costs of providing non-market benefits.

Forest Decentralization in Scotland

The push down of power

The push down of power over forests in Scotland is driven by the devolution of selected political powers by the UK government to the newly created Scottish Executive in the new Scottish Parliament and by the decentralization of UK forest management to the Scottish Forestry Commission. International commitments have played a role in influencing the decentralization agenda. In parallel with the changes in UK forest governance, the government has begun to address historical inequities, particularly with respect to land tenure.

Scotland has implemented the decentralization of power over natural resources primarily by making forest governance more participatory and by the land reform process.

Participatory forestry

The 1992 Earth Summit in Rio and the Council of Ministers meeting in Helsinki in 1993 focused attention on the non-timber benefits of forests, notably their 'environmental, cultural, spiritual, economic and social benefits' (UN, 1992). The UK government signed up to the Forest Principles and Agenda 21, thus committing to recognize the full range of forest benefits and to empower communities to enjoy them. In 1992, the government launched its *Rural Framework* for Scotland, in which it committed to 'returning power to the people' (Scottish Office, 1992).

In 1994 the UK Forestry Commission hosted a multi-stakeholder meeting on forests and people in rural areas and set up a forests and people in rural areas initiative. In 1996 it commissioned a study, *The Scope for Community Participation in Forest Management* (Slee et al, 1996). At the same time, the Forestry Commission began entering into formal agreements for co-management of some state forests with local communities.

In 1997, the New Labour party immediately began a political process to develop 'community planning' as a methodology, across all sectors, for involving communities in the decisions that affect them. Another immediate impact was a cessation of the privatization programme that had begun in 1989, with a moratorium on large-scale disposals of state forest.

Ownership of Scotland's national forests passed to Scottish ministers in the new Scottish Executive in 1999; but the forests were still managed by a centralized UK Forest Enterprise. After wide consultation, a new *Scottish Forestry Strategy* (Forestry Commission, 2000) was published in 2000. That same year, a consortium of agencies and non-governmental organizations (NGOs) published case studies of rural development forestry, along with a 'toolbox' for community participation (McPhillimy, 2000). Following one of its recommendations, the Forestry Commission set up a panel to advise on community participation in managing the Scottish national forest estate.

A major study on the level of community involvement within Forest Enterprise in Scotland followed (Hobley, 2002). Recognition of the importance of community involvement led to a programme of change within the agency, including staff training, seminars, networking, pilot studies and performance indicators. Each forest district now is required to develop a prioritized plan for involving communities using a 'menu' of options ranging from better consultation to ownership by communities.

In 2003, the management of the national forest estate was decentralized from the UK level with the creation of the Forestry Commission Scotland, which acts as the Scottish Executive's forestry department. A grant scheme was established to provide financial incentives to private landowners (including community land trusts) that provided incentives specifically aimed at community involvement.

In August 2003 the Scottish minister responsible for forests convened a review panel to review the long-term role of Scotland's national forest estate; the panel began consultations in December. Shortly, new local forestry fora would be established to advise on taking forward the Scottish forest strategy at regional levels, marking the start of a formal process of decentralization below the country level.

Land reform

One of the unique aspects of land tenure in Scotland is the system of 'crofting', under which the indigenous population in the Highlands and Islands of north-west Scotland were granted security of tenure after a widespread programme of clearing the people off the land in the early to mid 19th century. The 'clearances' were the forerunner of colonization and displacement, carried out on a global scale by the UK government for 100 years.

Crofting is a form of land tenure in which an individual has heritable rights to dwell on and manage a small area of land, called the croft or inbye, typically less than 10ha. Crofts were originally subsistence holdings, but have evolved into part-time agricultural units. The tenure arrangement defines a relationship between the crofter (tenant) and the owner of the land (landlord), in which both have rights and responsibilities towards each other and over the land. The arrangements were originally enshrined legally in the 1886 Crofting Act.

Crofting tenancies are organized into townships. In addition to their personal inbye land, most crofters also have a legal share in an area of common land, called the common grazing, which is attached to each township and is managed by a grazings committee elected by the crofting tenants. There are about 1000 common grazings across the Highlands and Islands. Typically, 15 to 20 crofters share an area of common grazing, 400ha to 500ha, which is usually hill land unsuitable for cultivation. Each area of common grazing has a set of rules that regulate land use – for example, the number of stock each individual crofter may graze on the land. There are around 17,000 crofting tenancies occupying 800,000ha, or 20 per cent of the Highlands and Islands, or almost 10 per cent of the total land area of Scotland.

As part of a political process to increase individual private ownership, the Crofting Act was reformed in 1976 to grant individual crofters the absolute right to buy their house site and garden ground and a qualified right to buy their inbye croft land. To this day, only 20 per cent of crofters have taken up this opportunity. Then in 1990, the government offered to transfer at no cost the ownership of its seven crofting estates to a single Scottish land trust that would be owned and managed by the crofting tenants. The crofters rejected the plan as having no advantages over state ownership.

Although crofters have rights to graze livestock on the common lands, they have no rights to exploit fish, game, minerals, water or other resources of the land and, until 1991, had no rights to manage woodlands. During the late 1980s, the Scottish Crofters Union (now the Scottish Crofting Foundation) advanced a convincing case for the benefits of crofter forestry on social, economic, agricultural and, not least, environmental grounds.

Between the two World Wars, considerable areas of common grazings had been taken over by the Forestry Commission to develop plantations with the promise of employment to crofting communities. But with increased mechanization and contract forestry, the number of local jobs had been steadily declining, and nationalized forestry was not able to deliver any social benefits to crofting communities. There was, however, confidence that locally managed crofter forestry schemes would be able to deliver social benefits such as employment and training – both in the short term from new plantings, fencing work and so on, and also for future generations. Scenic and amenity values could also produce economic benefits through tourism – an important source of revenue. Crofters also saw significant benefits to diversifying land use to include forestry, together with shelter, soil rehabilitation and fencing.

A notable aspect of the crofters' campaign was their emphasis on native woodland, in contrast to the predominantly exotic softwood forestry. The environmental benefits of crofter forestry include proper management of existing native woodland remnants on the grazings, many of which are ancient and of great ecological significance, and development of new native woodlands. The consequent benefits to wildlife, soils, water quality and fish resources ensured the cooperation of vocal environmental NGOs, such as the Royal Society for the Protection of Birds (RSPB), government agencies including Scottish Natural Heritage and the Scottish Landowners Federation (whose interests primarily are the 'sport' on their estate: both deer and fish would clearly benefit from native woodlands).

In 1991 the UK parliament passed the Crofter Forestry (Scotland) Act of 1991 to extend the powers of grazings committees by granting three new rights:

1 the right of any crofter to request their grazings committee to pursue forestry activity on the common grazings;
2 the right of the grazings committee, subject to the approval of the landlord, to 'plant trees on, and use as woodlands, any part of the common grazing', as long as 'not the whole of the common grazing is planted with trees and used as woodlands'; and
3 the right of the grazings committee to apply for grants for woodland management and afforestation.

Though falling short of actually granting ownership of the trees to crofters, this new law enabled crofting communities to become involved in afforestation and woodland management for the first time on a legal basis and to be able to share in the benefits of these activities. The primary financial benefit is in the form of government grants, without which crofter forestry would not be financially viable. These include establishment grants for new woodlands and annual premiums guaranteed for 15 years to compensate for any loss of income from grazing.

Since 1991, crofters have set up nearly 100 crofter forestry schemes involving 1700 individual crofters and covering 9000ha mostly with new native woodlands. The total establishment grant has been more than UK£7 million, with management grants of around UK£0.5 million per year.

In 1995, the secretary of state for Scotland supported a new model of ownership of croftland by locally based crofting land trusts involving all of the crofters within a community. The Crofters Commission also set up a service to provide advice and financial support for crofting communities seeking to set up crofting trusts. In 1997, this landownership model was enshrined in legislation with the Transfer of Crofting Estates (Scotland) Act, which allowed the transfer of state-owned croftlands to such crofting trusts. Government-sponsored advisers and a budget of UK£10 million to support community land purchases soon followed; but to this day, the legislation has never been used and the state still retains ownership of all its crofting estates.

The process of devolution involved significant commitments by the government to address historical inequities and anomalies and to modernize Scotland's land tenure system. Inequities included the ancient feudal system in which '*feu superiors*' could impose 'burdens' on land, such as restrictions on development, and retain rights to the land despite selling the title. In 2000, the Scottish Executive set up a committee to consult widely on land reform, and that same year, the feudal system of landownership was abolished in law.

The Land Reform (Scotland) Act of 2003 created the opportunity for all rural communities, not just crofting communities, to have 'first refusal' on the sale of any rural land, including woodland, and granted all crofting communities the absolute right to buy their croftlands on a collective basis at independent valuation even against the wishes of the landowner. This was termed a 'Mugabe-style land raid' by opponents to the reforms. The next step in the modernization process is a crofting reform bill that will, among other things, enable new crofts to be created.

The pull down of power

Through a popular people's movement, communities in Scotland are gaining greater control over their local natural resources. Injustice and the desire to put right past wrongs fuel this revolutionary activity and increase the motivation to bring about change. In many cases, historical and widely perceived injustices have been the driver of change: injustices including the inequitable land ownership pattern and historically forced relocations or 'clearances'; a history of bad landlords; and tax breaks for game-show hosts and other rich investors to drain ecologically precious peat land and plant non-native tree plantations. Six cases demonstrate the influence of revolutionary grassroots activities; for other examples, see McIntosh (2002).

Assynt

In 1989, at the height of a buoyant land market, the owner of the 9000ha North Assynt Estate in a remote corner of the north-west Highlands sold the largest part of his croft lands to a Swedish-owned property company for more than UK£1 million. Three years later, the company went bankrupt. In a now-weakened land market shaken by the collapse in the Lloyds insurance market, the selling agent decided to offer the estate for sale in seven lots in order to maximize its sale potential. The 120 crofters were incensed at the proposal to break up their land, which was described in the sales brochures not as croft land, but as 'sporting land', 'wildlife paradise' and 'where the people are perhaps alien'.

At a June 1992 meeting called by the Assynt branch of the Scottish Crofters Union, the crofters unanimously agreed to try to prevent the sale. The Assynt Crofters Trust was established as a company limited by guarantee – a legal entity that limits the financial liability of the members to a nominal sum – and the crofters made it clear that if anyone bought part of the estate, they would exercise their right to buy the land under the terms of the 1976 Crofters Act and nominate Assynt Crofters Trust to take the title. The individual crofters' tenancies would remain intact, but the crofters collectively would own the land. The risk to potential buyers was real. It would suppress interest, leaving the selling agent with a virtually unsaleable asset and the liquidators unable to realize the asset.

However, rather than test the interest of potential buyers, the crofters decided to buy the estate complete with its fishing, game, woodland and mineral rights and so become the first crofting community to have the control, the opportunity and the responsibility of managing their land. They decided to offer a fair price based on a realistic valuation, not one based on the whim of millionaires, which is the usual form of evaluation for highland estates.

Their high-profile campaign lasted nearly six months, during which the resolve of the crofters was tested in a battle of wits with the selling agent, the liquidators and the Swedish bank. The crofters received significant media interest, huge popular support and the backing of key agencies.

The people of Scotland were hungry for a revolution and land reform. People sang in the streets to raise money. They sent stamps and took up spontaneous collections ('whip-rounds') in pubs. Fund-raising *ceilidhs* (gatherings for singing, dancing and storytelling) were held up and down the country. Crofters and their families dug deep into their own pockets and negotiated the support of several agencies, which agreed to provide grants to assist with the purchase in exchange for promises of development and conservation, including significant new native woodlands. In December 1992 the crofters bought the estate for UK£300,000 – less than one third of the price paid by the Swedish speculators only three years before – and so became the first Scottish community in modern times to take full control and responsibility for managing their land and its resources.

Those events, told in detail by MacAskill (1999), inspired many other communities and brought land reform to the front of the national political agenda. Ten years later, great strides had been made in agriculture, tourism and housing and a new company established to explore opportunities in renewable energy. The leading national daily newspaper, the *Scotsman*, wrote, 'Ten years ago, a group of crofters in north Sutherland took control of the North Assynt Estate. It was a move described by some as bordering on the lunatic. But a decade later, it can be argued that the buy-out changed the shape of land ownership in Scotland forever.'

Treslaig

Treslaig is a small community of 11 crofts, reached by ferry, on Loch Linnhe in the west of the Scottish Highlands. Together with the neighbouring and even smaller township of Achaphubuil, the crofters of Treslaig wanted to buy a state-owned woodland of 66ha adjacent to their common grazings when the Forestry Commission programme of forest disposals was announced in 1989. When their interest was spurned, there ensued a five-year struggle to convince the government agencies that the crofters were credible as the future owners and managers. Meanwhile, the commission tried to sell the woodland to a national NGO, the Woodland Trust, without consulting the crofters, and then tried to give the woodland back to the descendants of the owner from whom it had purchased the land in 1961. In 1994, after five years of intense political lobbying, the crofting community finally succeeded in purchasing the woods.

The local institutional arrangement in this case is a company limited by shares, called Treslaig and Achaphubuil Crofters Woodland Trust. The members of the company each have shares in it corresponding to their crofts, which can be subdivided or amalgamated as the crofts change hands. The company makes decisions by one-person-one-vote regardless of the size of the member's share holding, thereby operating as a co-operative. The objective of the company is to maintain the ecological value of the woods, increase habitat diversity, encourage ecological regeneration and discourage non-native species.

The ecological orientation of the trust has had a strong influence on both government agencies and some of the large national environmental NGOs. Prior to their struggle, the government agency responsible for nature conservation (now called Scottish Natural Heritage) believed that only environmental NGOs and the state had the competence and experience to be environmentally friendly land managers. The agency's sponsorship was required for eligibility to purchase through the forest disposals programme, and the trust became the first community body to achieve this sponsorship. In this landmark case, the national conservation agency gave a clear signal that local communities were valid environmental managers. Scottish Natural Heritage has subsequently played a crucial and supportive role in many other community land buyouts.

Abriachan

Abriachan is a scattered rural community of about 120 people that sits high above the shores of Loch Ness, in the central Scottish Highlands. In 1998, Abriachan Forest became the largest community-owned forest in Scotland (although it has since been superseded by Culag; see below). It consists of a mixture of a commercial plantation of mostly exotic tree species, some remnant Scots pines and a substantial area of naturally regenerating open woodland. The community purchase came about as a response to a fear that it might lose its traditional access to the land. In 1990, Abriachan was included in the forest disposal programme, and community members said they wanted to retain access to the land, particularly the land that had been their traditional source of peat and wood for fuel. However, in 1995, when Abriachan Forest was actually put on the open market, there was no mention of community access in the sales particulars. The outraged community set about trying to buy the forest. In March 1998 they succeeded in buying a 534ha part of it.

Like the Assynt Crofters Trust, the Abriachan Forest Trust was established as a company limited by guarantee; but unlike both the Assynt and the Treslaig groups, membership is open to everyone in the community. The aims of the trust are environmental (to increase the diversity and amount of native species in the forest and restore wetlands for biodiversity), economic (harvesting commercial timber and encouraging use of the forest for tourism businesses) and, most notably, socio-cultural, with a strong emphasis on its historical and archaeological heritage and an educational component about forest use and community involvement in its management.

Abriachan has been an inspiration throughout Scotland for its inclusive and creative approach to forest management. Management has involved the whole community from schoolchildren to the oldest people in the village. It is an institutional model recommended for community landownership by the Community Land Unit and in the Land Reform legislation. Abriachan has also played a critical role in networking with other community forestry groups by hosting meetings. In addition, one of its founding directors became the first chair of the Scottish Community Woodlands Association when it was formed in 2002.

Laggan

Laggan is a small community in the central north-west Highlands in an area of intense forestry activity. In 1900 the population was 929; by 1990 it had sunk to 200. The surrounding 1400ha Strathmashie Forest was managed by the central Forestry Commission's Forest Enterprise and was planted mainly with Scots pine and Sitka spruce between 1955 and 1964.

During the 1990s, a few individuals identified Strathmashie Forest as a potential source of employment if managed and worked by local people. When rumors circulated that the forest was to be sold as part of the disposal programme, a team was assembled to look into the feasibility of community management of the forest, with a view to providing jobs that could stem further depopulation. The Laggan Forestry Initiative was set up.

From 1992 to 1996, Laggan Forestry Initiative sought to secure its objective of managing the forest. It worked to build local support through a series of community workshops and regular meetings. In the face of intransigence from the local forestry office, it lobbied not only Forestry Commission directors but also the Rural Affairs Department of the Scottish Office, the local member of parliament (and parliamentary candidates), various Highland councillors and the secretary of state. All of these activities attracted considerable interest from the media; but the Forestry Commission refused to consider any arrangement that would allow the local group to manage the forest.

Then suddenly in 1996, the secretary of state for Scotland visited Laggan and announced his support of the Laggan initiative. Later that year, the new Forestry Commission director-general personally handed Laggan Forestry Initiative a paper outlining seven options relating to Strathmashie: community ownership of land and timber; community ownership of land; community ownership of timber; community lease; use permits; partnership agreements; or an internal concordat.

Laggan Forestry Initiative's preferred option was to buy the forest; but its valuation at UK£1.75 million made that impossible. To explore the option of partnership, the Laggan/Forest Enterprise Working Group was created. After lengthy and sometimes acrimonious negotiations, a formal agreement was drawn up that would allow the community an active part in the management of the Strathmashie Forest. To be able to sign a legally binding document, Laggan Forestry Initiative transformed itself into a new legal entity – the Laggan Forest Trust – a company limited by guarantee without share capital.

The Laggan experience forced the Forestry Commission to have direct experience of co-management and thereby learn from it. Subsequent engagement with other communities was undoubtedly informed by the Laggan experience. Laggan also demonstrates that although political will is vital in pushing power down, institutional inertia can, nevertheless, be difficult to overcome.

Cairnhead

The village of Moniaive forms the heart of the parish of Glencairn in Dumfriesshire, southern Scotland. At the top of the glen is Cairnhead Forest, a 1347ha plantation of Sitka spruce planted by the Forestry Commission during the late 1970s. When the forest was put on the market in 1996 as part of the disposals programme, the local community became alarmed at the prospect of having little control over its management, the effects of timber harvesting, associated job opportunities and preservation of traditional rights of access. They were reluctant, however, to buy the forest and instead encouraged the Forestry Commission to retain ownership. In 1999 the local community signed a formal, legally binding concordat for co-management of the forest with Forest Enterprise.

The forest is managed by Forest Enterprise staff, as well as members of two community bodies: the Cairnhead Community Forest Trust (a company limited by guarantee with charitable status) and the Cairnhead Community Forest Members Association (an informal supporters' group that allows people from outside the community to play a role in forest management). The management is business oriented, with a focus on maximizing employment opportunities for local people and diversifying the economic opportunities generated by the forest and the land. There is also a strong conservation thrust towards biodiversity and the water catchment.

The case of Cairnhead is notable for being much less confrontational than that of Laggan. The legal framework of the concordat is now one of the menu of options that Forest Enterprise uses to explore community involvement in forest management elsewhere. Cairnhead built agency confidence that negotiations with communities could be constructive and fruitful without becoming battles.

Culag

The parish of Assynt is situated in the remote north-west corner of Scotland and has a population of around 1100. The principal village is Lochinver, with a population of about 600. Culag woods stands on the edge of Lochinver, between the village and the sea, and extends to 36ha. It is owned by Assynt Estate, which also owns the huge sporting estate that covers most of the parish (and was the owner of the North Assynt Estate).

Culag woods is stocked with conifers interspersed with native trees – rowan, aspen, oak, holly, hazel, downy birch and willow. The woods have long been valued by the local community because it gives the village some shelter and has pleasant walks that lead to a rocky shore with quiet coves. By 1990, however, neglect and mismanagement had considerably reduced its amenity value. The paths were overgrown and needed maintenance. Some areas posed a danger to the public because hasty abandonment of felling operations a few years before had left part of the woods susceptible to wind blow.

In 1992, the community council considered taking on the management of the woods. In order to prepare a practical plan that would convince the owners and possible funding agencies that it could be entrusted with the task, the council commissioned a brief report, which recommended developing the wood as a community woodland. The Assynt Estate and Assynt Community Council came to a verbal agreement that the community could manage the Culag woods.

The Culag Community Woodland Trust is a company limited by guarantee. Membership is open to all residents in the parish. In 1996, after three years of sensitive negotiations between landowner and community, the estate and the trust signed a 50-year lease that allows the trust to manage the wood as a community woodland. Since then, the trust has attracted the support of the Forestry Commission and other government agencies to restore the woodland, enhance its amenity value, offer training in forestry skills and create employment.

In 2000, an area of land 4.8 kilometres from the village came on the market. It extended to 1200ha, two-thirds of which had been recently planted as a new native woodland. The Culag Community Woodland Trust received funding from the government agencies to acquire the land and manage it as community woodland. It now has ambitious plans to build car parks, trails with all-abilities access and fishing facilities.

The Culag Community Woodland Trust broke new ground by leasing land from a private individual, thereby adding to the menu of options for achieving community involvement in the land. Its success in attracting support for its second land acquisition is unique in the community woodland movement so far and demonstrates the widespread support from the government agencies for community landownership and its benefits. The trust has demonstrated that as a community body builds its capacity, it can manage increasingly large land holdings.

Impacts of decentralization

The events in the case studies are relatively recent, and our account has focused on process and outcome rather than on social, economic or ecological impacts. However, some effects can be identified.

At the national level, these successes have stimulated other communities to enter into negotiations with the Forestry Commission Scotland and private landlords for a greater role in managing the local woodlands. Each success contributes to a growing pool of options for change and increasing confidence in the political will to create conditions for such changes to be replicated. To date, around 100 community woodland groups have formed, and a Community Woodland Association (www.community-woods.org.uk) has been established.

Within the communities, some antagonism has emerged between the crofters and those in the community without crofting status who have no say in managing the woodlands and do not share in any of the benefits. This is partly addressed in the Land Reform (Scotland) Act of 2003, which stipulates that any community land purchase made under the terms of the act must involve governance structures that are supported by the majority and are wholly inclusive.

The economic impacts of community ownership or management sharing are already becoming apparent and are now accepted by Forestry Commission Scotland as an incentive to community participation. The benefits include giving priority to local contractors, training local people in forest skills, opening up access to additional financial incentives for multipurpose forestry, such as access for recreation and health, and the creation of local project management jobs (Birnie, 2003).

Significant positive environmental impacts have arisen directly from community ownership and community partnership arrangements. Virtually all of the crofter forestry schemes have involved the creation of new native woodlands on degraded grazing lands. The other community woodlands have all involved better management practices and significant restructuring to decrease the amount of exotic species and increase the native woodland component.

LESSONS LEARNED

Revolutionary grassroots activity that pulls power down to the local level can shape decentralization processes. Communities are not passive recipients of power pushed down from the centre; rather, power shifts are evolutionary, involving push-me-pull-you interactions.

A timeline demonstrates clearly how the local struggles of communities both set the pace and set the agenda for land reform and community involvement in forest management in Scotland:

- 1886: Crofting Act gives security of tenure to crofters.
- 1976: Crofting Act reform gives crofters the right to buy their crofts.
- 1987: Highlands and Islands Forum conference on Land, Wildlife and the Community is held.
- 1989: Treslaig crofters seek to buy Forestry Commission land surrounding their crofts.
- 1990: Government crofting estates are offered at no cost to crofters, but are rejected.
- 1991: Crofter Forestry (Scotland) Act gives crofters the right to establish and manage woods.
- 1992: Assynt Crofters Trust takes over North Assynt Estate.
- 1992: UK signs Agenda 21 and Forest Principles at the Rio Earth Summit.
- 1994: Highlands and Islands Forum Community conference on The People and the Land is held.
- 1994: Grassroots Scottish Rural Development Forestry Programme is initiated.
- 1994: Treslaig becomes the first community to purchase forest from the Forestry Commission.
- 1994: Forests and People in Rural Areas Initiative is established.
- 1996: Culag arranges formal lease from a private owner.
- 1996: Forestry Commission report, *The Scope for Community Participation in Forest Management*, is published.
- 1997: Transfer of Crofting Estates (Scotland) Act is passed.
- 1997: Community Land Unit is set up.
- 1998: Abriachan becomes the last community to purchase forest from the Forestry Commission.
- 1998: Laggan agrees to formal community partnership with the Forestry Commission.

- 1999: Cairnhead agrees to a formal concordat between the community and the Forestry Commission.
- 1999: Ownership of Scotland's national forests devolves to Scottish ministers.
- 1999: Scottish Executive commits to land reform legislation.
- 2000: Culag purchases private forestland.
- 2000: Community Scottish Land Fund is established.
- 2000: Scottish Executive sets up Land Reform Group.
- 2000: Forestry for People Panel is established to advise the Forestry Commission on community participation.
- 2002: Scottish Community Woodland Association is formed.
- 2002: Forestry Commission report, *Community Involvement* (the 'Hobley' report), is published.
- 2003: Forestry Commission is asked to make forestland available for new crofting communities.
- 2003: Forestry Commission Scotland is created.
- 2003: Land Reform (Scotland) Act is passed by the Scottish Parliament.
- 2003: Scottish Executive announces setting up Local Forestry Fora.
- 2004: Scottish Executive sets up major review of national forest estate.

Local revolutions in Scotland influenced the push down of power by bringing about results that were perceived to be impossible and thereby firing the collective imagination of the country. They have provided concrete examples of how big problems (such as land inequity, bad land management, rural depopulation, loss of access to land, disempowered communities and unemployment in the forestry sector) can be solved. By achieving unexpected and welcome outcomes, such as educational innovations, community benefits, social housing, better environmental management, increased tourism and other economic gains, they have spread optimism, gained credibility and attracted media attention. The political will thus generated creates a positive feedback loop: politicians direct agencies to replicate the process, thereby reinforcing the push down of power.

Despite political will, there may well be institutional inertia, cultural resistance within agencies and lack of skills in participatory and locally driven decision-making. Agencies such as the Forestry Commission still have much to do in implementing decentralization, listening to local people and including community representatives on decision-making bodies. These problems require institutional solutions.

It is difficult to assess whether the Scottish experience is transferable to other peoples and cultures. The 'push-me-pull-you' of devolution of forest management in Scotland is taking place in a specific cultural and legal context: a post-industrial society where forestry is a small part of the gross domestic product, there are no forest-dependant communities, and forest management for timber and pulp is uneconomic and highly subsidized. There now is an increasing recognition that forests in Scotland will serve the national interest best if they are managed as multipurpose forests catering to health and recreation, as well as to environmental services and the production of timber. It is this shift in vision that is creating the space for community involvement in the management of the national forest estate and for community ownership – a space that many communities are now willing to occupy.

REFERENCES

Birnie, R. (2003) *The Social and Economic Impact of a Community Woodland*. BSc (Hons) dissertation, Edinburgh University, Edinburgh

Forestry Commission (2000) *Scottish Forestry Strategy*. HMSO, Edinburgh

Hobley, M. (2002) *Responding to the Challenge: Forest Enterprise and Community Involvement in Scotland*. Forestry Commission, Edinburgh

MacAskill, J. (1999) *We Have Won the Land: The Story of the Purchase by the Assynt Crofters' Trust of the North Lochinver Estate*. Acair, Stornoway, UK

McIntosh, A. (2001) *Soil and Soul*. Aurun Press, London

McPhillimy, D. (2000) *Rural Development Forestry*. Scottish Natural Heritage, Edinburgh

Scottish Office (1992) *A Rural Framework*. HMSO, Edinburgh

Slee, B., Clark, G. M. and Snowdon, P. (1996) *The Scope for Community Participation in Forest Management*. Forestry Commission, Edinburgh

The UK Government (1994) *Sustainable Forestry: The UK Programme*. HMSO, London

UN (United Nations) (1992) *Non-legally Binding Authoritative Statement of Principles for a Global Consensus on the Management, Conservation and Sustainable Development of All Types of Forests*. Report of the United Nations Conference on Environment and Development, Rio de Janeiro, 3–14 June, United Nations, New York (www.un.org/documents/ga/conf151/aconf15126-3annex3.htm)

Wightman, A. (1996) *Who Owns Scotland?* Canongate Books, Edinburgh

Wightman, A. (1999) *Scotland: Land and Power*. Luath Press, Edinburgh

WWF (World Wide Fund for Nature) and Forestry Commission (2000) *Protected Forest Areas in the UK*. Oxford Forestry Institute, Oxford

Chapter 13

Main Features of Russia's Forest Management System

Natalia V. Malysheva

Introduction

Historically, Russia has a very strong base for centralization but not for decentralization. The transition from a strongly centralized, multilevel system to a decentralized one is a distinguishing feature of forest management in Russia today. Management functions were separated from user-defined functions in forest administration after the downfall of the USSR. The government has reserved forest management functions for itself. User-defined functions are being transferred to the private sector, and administrative functions are being transferred from central federal bodies to regional and local ones. A new forest code is being considered by the Russian Parliament. Russia is seeking a better way of transferring the administrative, regulative and operational functions from the central government to regional or local forest management levels. The proposed forest code would strengthen the role of the regional executive bodies and transfer to the local governments broader authority regarding forest use, forest protection and reforestation.

Russia's forest management structure

Analyses of changes in the forest management structure and changes in the forest estate show that in Russia, centralization had a positive influence on the forest and utilization of forest resources. And vice versa: decentralization in forest management, in which power was transferred from the federal to regional levels and management functions were delivered to other authorities, resulted in forest exploitation and great economic and ecological losses.

All forests were nationalized and managed by local administrative bodies after the October 1917 Revolution. The forests were over-cut to supply the destroyed economy with wood and provide citizens with fuel. Later, during the 1920s and 1930s, all decision-making and control functions held by local administrative bodies were transferred to the central administration. The forest management units, as the representatives of the central forest administration at the local level, controlled forest use. The results of such control were positive for the forest estate.

The next attack on the forests was under Nikita Khrushchev's reforms of 1957–1966, when the economy was reorganized based on territorial divisions. Forest management and administration were transferred to the regions. The forest management units were joined together with state timber industrial enterprises. The forests were cut without control. Actual harvesting was many times above the reasonable volume of felled timber. Every fifth cubic metre of wood was harvested above the available standard (Petrov et al, 1997).

Nevertheless, Russia remains one of the major forested countries in the world. Forests cover about 800 million hectares (ha) of its territory, and more than 25 per cent of the global volume of standing forest is concentrated in Russia. Most Russian forests are located in the boreal zone; these forests represent about 60 per cent of the world's boreal forests. Many of these forests are characterized by low growth potential and high vulnerability, being extremely sensitive to any intervention. Russian forests are crucially important for the whole planet because they protect the environment, mitigate climate changes and serve as the world's largest sink for carbon storage.

In Russia, forest resources are associated with the concept of the Forest Fund. The term derives from the peculiarities of the history of the state's forest management. The Forest Fund encompasses more than forest, and historically, as civil and forest legislation developed, it became closer to the notion of real estate. Nonetheless, we will consider the Russian Forest Fund equivalent to 'forest estate' in English. The forest estate is managed for forestry purposes and comprises three types of land:

1 lands covered with forest and shrub vegetation (true forests);
2 lands that could be forest but are currently not forested (such as felling sites, burned areas and perished stands); and
3 non-forest lands (swamps, water, sands, roads, pastures, hayfields and others).

The forest estate has a permanent value and changes due to land transfers for industrial construction, farming or other agricultural use are insignificant. The total area of Forest Fund lands is 1,113.84 million hectares, or 69 per cent of the total land area of Russia (as of January 2002).

Generally, four main components constitute the system of forest management:

1 property rights on the forest estate and patterns of ownership;
2 the structure of forest management and the division of management functions among the forest management bodies;

3 the relationships between management bodies and forest users; and
4 finances for forest restoration, reforestation, forest protection, etc.

Property rights

According to the 1997 Forest Code of the Russian Federation, the forest estate and all forests except urban forests are under federal jurisdiction. The federal law allows the transfer of the forest estate to regions of the Russian Federation (Forest Code, Article 19). This means that the forest estate is in cooperative ownership of the federation and its regions. A new forest code is currently under development and will be brought to the *Duma* (the parliament) for discussion in 2005. A revised version presented for public discussion in the *Russian Forest Newspaper* (November 2004) provides for the preservation of state ownership of the forest estate. Nonetheless, civil legislation of the Russian Federation has established different ownership patterns, such as private, state, municipal and others. Both civil legislation and the proposed forest code guarantee free access to all forests. According to the current forest code, citizens and juridical bodies have the right to use the forest estate through concessions, long-term leases, short-term leases and free of charge. Most forest use is under short-term lease or free.

Structure of forest management

Russia has extensive forest management experience spanning more than 200 years. The Forest Department was set up in 1798 by an edict of Emperor Pavel I. The established structure of forest management has proved to be successful and the main principles have been kept to the present day. According to the current legislation, the state forest administration includes forest use monitoring and control activities, as well as protection and reforestation throughout the country.

Three main levels are represented in the Russian forest management structure: federal, regional and local (forest management units). Administrative and management functions are carried out by the government of the Russian Federation, executive bodies of the administrative units of the Russian Federation, and specially authorized state forest administration bodies. Such bodies are represented by the Ministry of Natural Resources, which governs and controls 96 per cent of Russia's forest estate – almost 69 per cent of Russia's total land area. The ministries of agriculture, education and defence manage the rest. The minister of natural resources is nominated by the president of the Russian Federation.

Administrative reforms now taking effect will establish a new structure of forest management. According to a March 2004 president's decree, the federal level of forest administration is represented by three specially authorized bodies responsible for:

- legislative regulation of forest use, forest protection and reforestation;
- forest estate management; and
- implementation of ecological and forest legislation.

According to the 1997 version of the forest code, the regions have the right to own the forest estate and use the forests in administrative units; license short-term leases (up to five years), long-term leases (up to 49 years) and use forestlands free of charge; and determine the rental charges and forest rates over administrative units. The new proposed forest code strengthens the role of the regional executive bodies in forest administration and management by giving them the right to make arrangements for forests in the regions.

The new forest code also institutionalizes broader authority for local governmental bodies regarding forests growing on municipal lands. They will be responsible for determining the order of forest use and forest protection on municipal lands; participating in the selection of forest areas intended for different activities over administrative units; fixing stumpage payments for the local forests; and making agreements dealing with forest stands offered for sale, etc.

Until very recently, Russia did not have a political system that involved people in decision-making or permitted participation in forestry projects at the local level. The rights being transferred to local governmental bodies are intended to be the first step towards achieving improved forest management at the local level.

Relationships between managers and users

Under the Soviet system, all types of forest management and activities – administration, making arrangements for the forest estate, use-defined issues, regeneration activities and control functions – were covered by forest management units in the centralized planned economy. Central planning government agencies carried out all administrative functions. The forest management units were only executors of planned tasks and were prevented from making their own decisions. The effectiveness of control was quite low. From one side, the forest management units involved in timber harvesting, together with state timber industrial enterprises, controlled themselves. Yet, they also controlled the activity of state timber enterprises executing planned targets developed by central bodies at higher levels of administration. Thus, forest management units as state enterprises were not able to fulfill management functions together with the administration. A legal framework for a relationship between forest users and forest managers did not exist.

The decision to separate management functions from user-defined functions in the administration of forests was made after the USSR's demise. The primary forest use is wood harvesting by final felling and selective logging. The government relinquished the user-defined issues but reserved the forest management functions for itself. User-defined issues were delivered to the private sector because of a ban imposed by legislation on wood harvesting and timber reprocessing by state bodies of the forest service. Previously, the state entities had been active participants in wood harvesting and reprocessing; but beginning in 1993, they were limited to management and administrative functions. Later, the forest service, devoid of income from final felling, reserved the right to conduct selective logging during the transition period.

Ninety-eight per cent of Russia's timber industry was privatized. Naturally, the new private sector took advantage of the great differences between export and domestic prices on timber products. Without a clear customs and tax collection system during the mid 1990s, an unusually high profitability for roundwood and timber products export followed from those differences. Forest exporters and their partners profited handsomely as a result of the liberalization of trade. By leaving the timber market, the government has, in fact, lost a substantial part of forest income.

Throughout Russian history, forests had been under centralized control. Management agencies had high status in the state system and sufficient power to use forest income for the country as a whole. Suddenly, the state forest management system had no forest income and became fully dependent upon the federal budget. Limited budgets led to a weakening of control over forest utilization and the forest estate. The local forestry administrations, meanwhile, having reserved the right to cut during the transition, started to sell harvested wood and earned money from the sale of tax-free timber. Replenishing internal funds this way was severely criticized by Russian forest economists and Russian non-governmental organizations (NGOs).

A main feature of forest management and administration in Russia today is decentralization in decision-making. The core of the forest management and administration structure is still the forest management unit; but now centralization has been replaced by a very high degree of decentralization in forest management. User-defined issues such as final felling and timber reprocessing have been transferred to the private sector. Both decision-making and management of the forest estate have been given to the regions of the federation. Forest management units are now responsible only for forest estate control and reforestation. This means a positive and significant change in the status of forest management units, which now have management functions and restricted operational activity. A legal relationship between management bodies and forest users is realized through agreements in the form of a lease or forest auction.

Financial maintenance

From the 1930s to the 1990s, the finance system was based on state budget financing and distribution of the budget to the regions and forest management units. Payments for forest resources were requisitioned and added to the national budget. Financing of forest management bodies did not depend upon payments received from forest harvesting. With the 1997 Forest Code, sustainable forest management in Russia is designed to solve environmental and economic problems based on important local, regional, national and global issues. The financial structure has to ensure that the means and resources for reforestation and maintenance of forest protection are guaranteed, and that incentives exist to invest and raise revenues needed for all aspects related to the forest sector.

The financial system relies on getting forest revenues through a system of payments for use of the forest estate. Under Russia's market economy, such a system has become an effective instrument for managing the economic and legal aspects of forest use. The system of payments is of great importance for promoting sustainable use and reforestation of the land. Timber harvesting is still the main source of revenue.

The establishment of reasonable and well-grounded rates on forest tax and non-timber values of forests has been proposed as a fundamental measure to ensure revenue and replenishment of the budget. According to the 1997 Forest Code, the payments for use of the forest estate are exacted as forest taxes and payments for leases. Forest tax rates are established by administrative bodies of the regions of the Russian Federation as a result of lease agreements or forest auctions. The minimum stumpage payments are approved by the central government.

According to the forest code, forests are under state ownership. Major expenses of state forest management, such as forest guards, forest protection, reforestation and sustainable use, have to be covered by the federal budget, and therefore the state has the right to receive revenues from Forest Fund utilization. The following activities and expenses are financed by the federal budget:

- federal forest administration;
- regional forest management;
- national parks administration;
- forest fire protection;
- protection of forests against pests and diseases;
- forest reclamation;
- construction, maintenance and repair of irrigation networks;
- seed breeding;
- forest monitoring; and
- upkeep of the Forest State Account, State Forest Cadastre, etc.

Although the forest code stipulates that forest regeneration costs must be covered by the budgets of the regions of the Russian Federation, this does not occur in practice. Reforestation in several administrative units is not financed by the regions and is not even foreseen in regional budgets. For example, in 2001, only 11 per cent of the money needed for reforestation was provided by regional and municipal administrations (Roshchupkin, 2003). Thus, reforestation was really financed by the forest management units from their own internal funds.

Currently, the structure of financial support for forest management consists of the following:

- The federal budget covers management activities, forest fire protection activities and forest protection from pests and diseases.
- Funds from regional budgets of the Russian Federation are allocated for forest regeneration.
- Forest management units obtain additional revenue from rental charges, forest taxes, various earnings from the sale of forest products, gains from shelterbelt afforestation, etc.

It is interesting to compare the dynamics and structure of expenditures on forest management before the forest code was enacted and now, ten years later. In 1992, the federal budget accounted for 70.8 per cent of forest management expenditures, compared with 24.2 per cent in 2001. The forest management units generated only 20.4 per cent of their own budgets in 1992, versus 65.5 per cent in 2001. The amount contributed by regional administrative units has changed less: 8.8 per cent in 1992 and 10.3 per cent in 2001.

In 2002, forest management expenditures were divided as follows: the federal budget covered 34.5 per cent of the needed sum, the budget of the regions of the Russian Federation covered 8.5 per cent, and internal funds of the forest management units amounted to 57 per cent (Roshchupkin, 2003). This means that the federal budget is nowhere near enough to cover forestry expenses; furthermore, state financing is gradually decreasing. The shortfall from the federal budget was supplemented with additional finances earned by the forest management units. Internal funds of the forest management units have also been supplemented with revenue from forest products sales, primarily from selective logging, but also from transport services payments, shelterbelt afforestation, seed and planting stock sales, and other revenue sources. From year to year, the forest management units have been forced to increase selective logging and other profitable businesses to support necessary operational activities. Thus, budgetary shortfalls are primarily met through timber sales by forest management units.

Restricted financing by the state budget strongly affects federal properties and limits the capacity of state agencies responsible for forest fire protection, pest and diseases protection, research institutes and others. For example, the decrease in funds for forest fire protection led to the increase in disastrous forest fires during the past few years (Ministry of Natural Resources, 2002).

Changes in the rules of financing for forestry have been observed over recent years (Roshchupkin, 2003). For example, in 2002, all the payments based on the minimum stumpage cost were transferred to the budgets of the regions of the Russian Federation as revenue. The federal budget received the difference between the sum of rental charges and other forest payments and the total value of the minimum stumpage costs. Earlier, the difference was transferred to the forest management units. Despite these changes in the rules, federal budget contributions to forestry financing are decreasing, and expenditures from the federal budget exceed revenues.

The financial system, on the whole, is the same as in the former centrally planned economy. Forest management agencies cannot collect and manage the forest revenue themselves, but continue to get their funding from the state budget.

Conclusion

Decentralization as a basis for forestry administration that is more efficient and more responsive at regional and local levels is a crucial issue for Russia. Centralization is being replaced by a very high degree of decentralization in forest management. User-defined functions are being transferred to the private sector and arrangements for the forest estate are being transferred to the regions and administrative units of the Russian Federation, which are ill prepared to fulfill these functions.

The development and implementation of decentralization principles will succeed only if authorized regional agencies with regulatory functions are capable of making decisions and being responsible for those decisions. The success of decentralization also depends upon the availability of sufficient resources and the ability to use these resources autonomously. This requires local administrative bodies to collect local taxes and fees and to procure enough funds from the central government to execute the tasks transferred to them. However, since the finance system in Russia is the same as in the former centrally planned economy, forest management units cannot collect and manage forest revenue themselves, but must rely upon money from the state budget. This means that the financial and administrative systems for forest management are not adequate for rapid decentralization.

Too-rapid steps towards decentralization without efficient management and a legally accountable administration may have serious consequences. The history of forest resources in Russia demonstrates that better management of forests is not an automatic result of decentralization. Today, Russia looks for a reasonable distribution of administrative, regulative and operational functions between different forest management levels. The proposed forest code seeks to strengthen the role of the regional executive bodies and transfers to local governments broader authority over forest use, forest protection and reforestation. The right of local people to participate in decision-making procedures through public discussions should promote better and more transparent forest management and the implementation of democratic principles.

Sustainable forest management is a priority for Russia and is reflected in the new legislation. While maintaining a multilevel structure for forest management, the proposed forest code's strengthening of the federal hand to implement forest legislation is an indispensable condition for successful forest management in the future. A strong legal framework that sets out the rights and duties of all administrative levels will clearly provide good conditions for better management of the forest estate.

References

Ministry of Natural Resources (2002) *Russia's Forests*. State Forest Service, VNIILM, Moscow

Petrov, A., Mamajev, B., Tepljakov, V. and Shchetinsky, E. (1997) *State Forestry Management*. ARICFR, Moscow (in Russian)

Roshchupkin, V. P. (2003) *State Report about the State of the Forests and Use of Forest Resources in the Russian Federation in 2002*. VNIILM, Moscow (in Russian)

Chapter 14

Forest Resources Decision-Making in the US

Gerald A. Rose, with Douglas W. MacCleery, Ted L. Lorensen, Gary Lettman, David C. Zumeta, Mike Carroll, Timothy C. Boyce and Bruce Springer

INTRODUCTION

The US is an example of a country that started with a highly decentralized political and economic structure and has, over the decades, centralized portions of that structure, while leaving other portions decentralized. US forest ownership is diverse and includes extensive private forests, federally managed forests and public forests managed by state and local governments. Although a few states have laws that regulate forest practices on private land, most rely upon voluntary best management practices and technical assistance. All private forests are governed by laws relating to the protection of water quality, wetlands and endangered species; but private forest owners' objectives guide management for other values. Educational and technical assistance are provided by the state and federal government, with incentives to meet priority public objectives. Stakeholder involvement is considered important at all levels of government and is encouraged through specific provisions. Although there is significant disagreement on the meaning and indicators of sustainability, there is broad agreement on the importance of managing US forests well for current and future generations and a willingness to work together.

THE US FOREST ESTATE

The US has the fourth largest existing forest estate of any nation, with 8 per cent of the world's forests, exceeded only by the Russian Federation, Brazil and Canada. About 33 per cent of the land area, or 302 million hectares (ha), is forested. These forests and woodlands vary from sparse, scrub woodlands of the arid, interior west to the highly productive forests of the Pacific Coast and south, and range from pure coniferous forests to multi-species deciduous forests.

About two-thirds (204 million hectares) of the nation's forests are classed as productive forests that are not legally withheld from timber harvest. About 7 per cent (21 million hectares) are reserved for non-timber uses and are managed by public agencies as parks, wilderness or similar areas. The remaining 26 per cent (77 million hectares) are unreserved but considered unproductive for growing industrial wood; they are important for soil and watershed protection, biodiversity and wildlife habitat, livestock grazing and other uses.

Private forests account for 58 per cent of all US forestland, federal lands about 33 per cent, and state and local governments the remainder. Forestlands managed by industrial forest products companies constitute 13 per cent of productive, non-reserved forestlands and are generally concentrated in the south and the Pacific Coast. Forests managed by non-industrial private and family forest landowners are 58 per cent of the productive non-reserved forestland and are primarily concentrated in the east. Federal lands are concentrated in the west and total 261 million hectares, or about 29 per cent of the land area. State and local governments manage 7 per cent of US productive non-reserved forestland (USDA Forest Service, 2001). State and local governments also manage productive reserved forests and unproductive forestlands, such as parks, recreation areas, wilderness and special protected areas (MacCleery, 2002).

Early in US history, forests were used for fuel, building materials and special products such as ship masts. Otherwise, forests generally were cleared for growing food crops for people and domestic animals. As the population grew, the demand for timber for building materials increased tremendously. As the forests were thought to be inexhaustible, harvesting increased rapidly, leading to a crisis in supplies after only a few decades. Public concern prompted responses through law, education and incentives to reforest the vast areas that had been cut. Eventually, the concept of forest sustainability merged with the early focus on timber. The realization that there was more to forest sustainability than just timber was reinforced by research and changing public values. Today, economic, ecological and social issues are important aspects of US forest management.

Evolution of Management Systems

At the time of their independence from England, the 13 original American colonies entered the union as largely autonomous entities, or states, and over the years they have guarded that status zealously. Under the US Constitution, the federal government has limited authority and responsibility; all other powers are reserved for the states. Forestland management and use was one such reserved power. Nevertheless, the trend over the decades has been to centralize some policy and institutional functions at the national or federal level.

Throughout the 19th century, rapid settlement and economic development of the public domain, especially the western lands, was encouraged. A variety of approaches were taken, including transfers of federal lands to individual farmers, ranchers and corporations, especially railroad companies. After 1850, the western lands began to be settled at an accelerating pace. Concerns began to be voiced over some of the environmental and economic implications of rapid development.

In response, policy and institutional changes were put in place during the early decades of the 20th century:

- closing the public domain to further conversion to private land and to reserving remaining public lands as national forests, national parks and national wildlife refuges;
- promoting and encouraging the protection of forests and grasslands, regardless of their ownership, from wildfire, insects and disease;
- improving the art and science of natural resources management and promoting the more efficient utilization of raw materials;
- improving the management and productivity of both agricultural lands and forests through technical assistance to farmers and landowners; and
- adopting and enforcing federal and state wildlife conservation laws.

The policy changes led to the centralization of some aspects of US forest policy at the federal level, including policies governing management of federal lands; some national policies and standards for institutional capacity-building at the state and local levels, especially for emergency wildfire protection; development of a national programme for forest inventory; federal policies governing management of migratory birds; and federal funding and priority-setting for research (MacCleery, 2002). Yet other aspects of conservation policy remained the prerogative of state and local governments and individuals. These included the regulation of private forest management practices, much private forest investment, state-directed research and forestry education at state universities and colleges (Steen, 1976).

Environmental laws passed in later years promulgated mandatory federal regulations directly affecting the private sector, such as federal standards for controlling air and water pollution; federal regulations for pesticide use; and limitations on private landowner activities that would affect endangered species or that would drain or fill a wetland (MacCleery, 2002). Laws also authorized citizens, non-governmental organizations (NGOs) and others to bring lawsuits to require federal agencies to enforce the laws, and even included federal financing of successful lawsuits against federal agencies. These provisions have substantially increased the role of citizens, NGOs and industry in implementing and enforcing environmental laws and the role of US courts in interpreting the 'intent of congress', and thus setting environmental policy through judicial interpretation. However, these laws have also resulted in a significant cost to the government, as well as delays contributing to similarly high costs to the private sector.

Over the years, increasing concerns have emerged over whether lack of coordination at the federal level impedes the implementation of environmental laws. Furthermore, during recent years both substantive and process requirements have increased for forest landowners and managers to meet federal, state and local environmental regulations. Some actors are concerned that redundant and duplicative process requirements imposed by federal and state regulatory agencies upon land management agencies, state governments and private entities are onerous (USDA Forest Service, 2002b). Others suggest that the processes are appropriate institutional adaptations to respond to new needs.

On private lands, the nature and amount of stakeholder involvement are the prerogative of the landowner or managing entity. Stakeholder involvement on private lands had been non-existent, except in special cases, until the advent of forest certification. With forest certification, most systems have strong stakeholder involvement in the development of certification standards. Private landowners, to be certified, must implement and follow the standards of the system and undergo periodic on-the-ground audits. Most private forestlands owned and managed by large forest industries in the US are certified or being certified; but only a small percentage of US family or non-industrial private forests are certified.

Current opportunities and challenges

The US has seen a pattern of stable or increasing forest area and biomass per hectare during the 20th century. Forest growth nationally has exceeded harvest since the 1940s, and the average standing volume of wood per hectare is about one third greater today than it was in 1952. The reasons include rising agricultural productivity with a consequent decrease in land conversion to agriculture; stable land tenure and resource use rights; consistent markets providing profit and value for forest-related uses; commitments to research, extension, inventory and coordinated response systems for forest management; strong economic performance allowing investments; and consolidation of public values supporting non-timber and environmental benefits of forests (MacCleery, 2002).

Current forest management issues include forest loss and fragmentation due to urban and suburban expansion; reduction and fragmentation of late successional and old-growth forest habitats because of timber harvesting; loss and degradation of riparian and wetland forest habitats; loss and deterioration of forest and grassland habitats once created by frequent low-intensity fire; the effects of air pollution on forests in some areas; and displacement of native species by introduced exotics, to name a few. Of particular concern are rare and unique ecosystem types and species with specialized habitat requirements that are associated with them.

Addressing these issues requires the US forest sector to grapple with critical tensions inherent to growth, development and sustainability:

balancing urban and rural interests, needs and values; forging a social consensus on the use and management of both private and public lands; designing effective incentives to encourage acceptable management strategies in both the public and private sectors; reconciling resource consumption with conservation needs; and understanding and finding solutions to growing concerns, such as invasive species and climate change.

The importance of community involvement and support

> *A crucial aspect of achieving sustainable communities at the local, regional and national levels is the involvement of all stakeholders from the inception of the policy making process (National Council for Science and the Environment, 2001).*

Communities of interest include NGOs, associations, industries and other organized groups. Their involvement in forest issues at the federal level and, to a lesser degree, at the state level is very important. Although many communities of interest recognize the importance of all three components of sustainability – environmental, social and economic – and work effectively together, others centre strongly on one of the components. Some NGOs seem to thrive on conflict rather than solutions, and stakeholder input often exposes disagreements among groups (including with and among government agencies) over management strategies, frequently culminating in lawsuits that greatly increase the cost of government and business without clearly furthering sustainability.

Whereas communities of interest, especially the environmental NGOs and the major forest industry associations, tend to be very well organized around forestry issues at the federal level and have political influence, communities of place generally have not been effectively involved in forest resource decision-making in the US, least of all at the federal level. Communities of place are organized around education, human health, transportation and similar common issues nationwide. Because of this, they can have difficulty addressing forestry issues, which differ regionally and over time. The major differences centre on forestland ownership and the degree of reliance upon forest-based employment. A major effort is needed to design and conduct more localized forums for citizen involvement. In addition, communities of place need assurance that their involvement receives strong consideration in decision-making on a par with that of communities of interest.

One size does not fit all

The US constitution limits the authority of the federal government to issues of national significance; powers not expressly covered at the federal level are left to the states. Therefore, each state has developed its own constitution to affirm and organize the powers left to the states. How each state conducts government operations varies considerably from state to state. In addition, there are social and cultural differences between different regions and states. Therefore, what works well in one region or one state may not be acceptable or workable in another. Each state has the freedom to adopt approaches to natural resource decision-making that work best for it.

Because forest resource decision-making not expressly covered in federal statutes is left to the discretion of state and local governments, and each state approaches forest resource decision-making in a different way, the institutions, approaches and degrees of decentralization in forest management differ among the states. The following case examples provide a good cross-section of the approaches to forest resource decision-making used in the US.

National forest system

Federally owned and managed forestland illustrates the trend for increasing centralization of management over the past 30 years, but also shows how federal agencies have benefited from public input. The National Forest System encompasses 77 million hectares, or 8 per cent, of the land area of the US and 20 per cent of its forestland (59.4 million hectares), most in the west (USDA Forest Service, 2001). The public lands were established at the beginning of the 20th century on the widespread assumption that public ownership of forests was needed to secure long-term supplies of timber. Contrary to that assumption, today 89 per cent of the timber produced in the US is harvested from private lands.

Dominant public values and intense public debate have always affected how these lands are managed. Especially since 1970, these debates have become ever more intense and polarized, reflecting public dissension over the overall mission that should govern these lands. This lack of agreement, coupled with implementation of federal environmental laws enacted since 1970, has resulted in a substantial shift in management focus, with increased emphasis on managing for amenity values and restoring ecosystem functions, biodiversity and health, and a corresponding reduction in timber and other commodity outputs.

The environmental laws of the 1970s, particularly the National Environmental Policy Act of 1970 and the National Forest Management Act of 1976, mandated that federal land managing agencies substantially increase public participation and encourage more open and 'transparent' consideration and evaluation of management alternatives. In essence, the forest service has shifted from being a mediator (receiving public input and deciding how to evaluate it when coming to a decision) to being a facilitator or collaborator (encouraging more interactive involvement among and between participants, other federal agencies and the forest service).

The new role places greater emphasis on effective collaborative skills in dealing with the public and other public agencies and relies heavily upon forging partnerships to carry out some of the tasks traditionally performed directly by the agency itself. There has also been an increasing focus on community-based participation and input. The result is a substantially reduced decision space available to field managers to respond to local demands and preferences. In particular, the shift in power from local communities whose economies were heavily dependent upon national forest commodity production to national and regional special-interest groups has left many of these communities with a sense of powerlessness about their ability to influence their own future (Lee, 1994).

The mission shift on National Forest System lands has coincided with growing concerns over ecological health. Significant areas are becoming unnaturally dense and unhealthy, putting both human and natural communities at risk to wildfire. It has been estimated that 16 million hectares of national forestlands are at serious risk of damage from catastrophic fire, insects and disease. The extensive and destructive wildfires of recent years have only underscored these concerns (GAO, 1999). For example, during the 2002 fire season, more than 2.9 million hectares burned, 21 firefighters were killed, tens of thousands of people were driven from their homes and more than 2000 buildings were destroyed.

Fuel build-up aggravates other problems affecting forest health. Thinning, other silvicultural practices and prescribed fire are seen as major, even essential, management tools to address these growing forest health problems. Yet, there remains a lack of public consensus about the national forest health problem and how to address it. Even relatively modest projects designed to test and evaluate various forest health treatments at the field level can be controversial.

Social consensus on how national forests should be managed is elusive. In grappling with the challenge of seeking such agreement, both the legislative and executive branches of the federal government have tended to create additional procedures and processes to ensure public awareness and participation. The forest service, other federal agencies, congress and the courts have added multiple layers to national forest planning and decision-making. Consequently, national forest management has become increasingly costly and time consuming. With considerable opportunity for individuals and interest groups to intervene to delay or block proposed actions, 'process gridlock' has occurred repeatedly.

A report by the US Department of Agriculture (USDA) Forest Service (2002b) contains a detailed discussion of how process requirements are adversely affecting the agency. It found that overlapping procedural requirements, procedural redundancy and multiple layers of inter-agency coordination contribute to inefficiencies in decision-making.

Moreover, although the reduction in national forest timber harvest levels has been justified on environmental grounds, public attention has not focused upon the environmental effects of the shift to forests elsewhere.

Today the US population consumes more timber and other natural resources than at any time in its history and also consumes more per capita than any other nation (US Bureau of the Census, 1975, 1994). The reduction in national forest timber harvest levels has acted to shift timber harvesting and its environmental and ecological impacts elsewhere – both in other countries and in other forest holdings in the US. Increased harvesting in the south-eastern US to supply chip mills, for example, has generated public concern over possible regional degradation.

The broad mandate for managing National Forest System lands was seen as an advantage when demands on these lands were relatively modest. During the early years, decision-making power over these lands was relatively decentralized, with local managers having considerable autonomy to make resource management decisions. As demands increased and the conflicts between alternative land uses became more apparent, policy and decision-making power became increasingly concentrated in the nation's capital. With federal lands, whose goals and stewardship must be accountable to civic constituencies, diverse constituencies seek a stake in how these resources are managed and have, consequently, organized to influence actions and policies. Given the concentration in Washington of authority for public land management, special interest groups often centre their resources in the capital as well, where constituencies and agencies may be disconnected from the economic and social impacts of policy shifts, such as reducing commodity production on public lands (Sedjo, 2001). In fact, the strength of such groups is in large measure responsible for those changes (Fretwell, 2001). Although there have been some modest attempts during recent years to give local communities more influence on decisions affecting national forest lands, the outcome is still far from certain.

Oregon

Oregon takes a centralized approach to forest management, using forest practice regulation with options and flexibility. The state plays an active role, with input and participation by stakeholder groups, in defining standards and monitoring performance by all classes of forest owners.

Oregon has tremendously diverse forests because of a wide range of climatic and geological conditions. Climate cycles, forest fires, windstorms, landslides, floods and insect and disease outbreaks are normal events in this dynamic landscape. Forty-five per cent, or 11 million hectares, of Oregon is forestland. Fifty-seven per cent of that (6.3 million hectares) is owned by the federal government. Private forestlands cover 4.5 million hectares; state ownership, 365,000ha; county and municipal ownership, 50,000ha; and tribal ownership, almost 195,000ha. Under private ownership, 2.4 million hectares are industrial and 2 million hectares are non-industrial (family forests).

Each ownership class in Oregon has a range of mechanisms designed to protect forestland for forest uses, water quality and diverse plant and animal species. The protections vary with the primary purpose of the lands (based upon ownership and zoning) and landowner objectives. All forestlands in Oregon are required to comply with state water quality standards and state land-use laws. They also must meet federal standards and regulations regarding endangered species.

The objectives of Oregon's land-use programme are to conserve forestlands by maintaining the forestland base and to protect the state's forest economy by promoting economically efficient forest practices that ensure the continuous growth and harvest of trees as the leading use on forestland. Economic production should be consistent with sound management of soil, air, water, and fish and wildlife resources and provide for recreational opportunities and agriculture. Lands zoned as forestland include lands that are suitable for commercial forest uses, operations and practices, and other forested lands that maintain soil, air, water, fish and wildlife resources, agricultural use and recreation. In these areas, residential uses and dwellings are significantly limited and must be sited in order to limit conflicts with forest uses and minimize wildfire hazards and risks. This programme is implemented through local government with oversight by state agencies.

Since its inception, the Forest Practices Act has been viewed as a partnership between the department and the regulated community. The best management practices have evolved substantially as new science and monitoring have identified necessary improvements. A monitoring programme determines the effectiveness of the best management practices and rules, assesses levels of compliance, and recommends modifications or additional rules.

A seven-member citizen Board of Forestry leads Oregon in implementing policies and programmes that promote environmentally, economically and socially sustainable management of both public and private forests. The 2003 Forestry Program for Oregon adapts an internationally recognized sustainable forest management framework – based upon the Montreal Process Criteria and Indicators – for use in discussing and measuring forest issues at the statewide level. This programme supports cooperation and incentives as the preferred tools for promoting desired public benefits on private lands. Oregon continues its strong regulatory approach through the state's Forest Practices Act and the Statewide Land Use Program; but it recognizes that additional public benefits are better provided through incentives.

The seven members of the citizen Board of Forestry are appointed by the governor and confirmed by the state senate. No more than three members may receive any significant portion of their income from the forest products industry. The board holds three mandated public meetings per year and generally conducts at least three additional meetings. Public participation is encouraged. The board is charged with supervising all matters of forest policy within Oregon, appointing the state forester, adopting rules regulating forest practices, and providing general supervision of the Oregon Department of Forestry.

By statute, the board and department of forestry are required to support standing public advisory committees: three regional forest practices advisory committees (on the operational and technical feasibility of proposed rules), the Committee of Family Forestlands and the Forest Trust Lands Advisory Committee (on state-owned forest management). In addition, the board and department routinely form advisory committees for specific tasks or issues. A range of NGOs in urban Oregon provide input to the board as it develops tools and approaches to 'operational sustainability'. These groups are taking a collaborative approach with local industries to promote sustainability.

Forest industry and family forest landowners are increasingly joining certification systems. A majority of the industrial lands belong to at least one certification system. The Oregon government has viewed these systems as market-based; as these systems mature, a role for government in supporting some of these systems may be considered.

A challenge to forest management in Oregon is the lack of public consensus for a forest management approach that balances and integrates conservation and wood production priorities in alignment with the goals of private, state and federal landowners. This has resulted in ongoing allocation of resources to address conflicts and failure to implement plans as intended.

The economic contribution of Oregon's forests has declined during recent decades, primarily because of reduced timber harvest from federal land. Harvest levels on private land have remained relatively stable for several decades. Employment in the forest sector has fallen during this period, creating significant economic challenges for rural communities. Private forestlands need to remain economically viable or pressure will mount to convert them to less environmentally friendly uses, such as urban development. Global forest product market forces and more lucrative alternative investments increasingly challenge private landowners. Oregon's high regulatory standards place the state at an economic disadvantage – unless forest owners can capture value for those higher standards.

The combination of fire suppression and inadequate forest management has left up to 75 per cent of Oregon's federal forest at moderate to high risk of uncharacteristic fire. Thinning and other management activities needed in a long-term programme to restore healthy conditions in federal forests will not only reduce fire risk and create jobs in rural communities, but will also help to protect forest ecosystems, including old-growth habitat, over the long term.

Minnesota

Minnesota has a legal and enabling framework that stresses commitment by stakeholders to work together to develop and implement sustainable practices voluntarily. This approach tries to balance centralized processes with significant decision-making by civic and private groups.

Extensive logging and catastrophic fires during the late 1800s and early 1900s drastically reduced Minnesota's forested area (Jaakko Pöyhry

Consulting, 1992). Settlement during the early 1900s resulted in large transfers of public domain and Native American lands to private owners. Widespread farm abandonment and large-scale tax forfeiture, especially in the 1930s and 1940s, led to reversion of about 3 million hectares to forest by 1953 (USDA Forest Service, 2002a).

Since 1977, Minnesota forestland area has been fairly stable at about 6.7 million hectares. Fifty-seven per cent is publicly owned – more than any state in the eastern half of the US (USDA Forest Service, 2002a). Minnesota administers 26 per cent of the total forestland; the federal government, 18 per cent; and county governments, 13 per cent. Private individuals and corporations own 43 per cent of Minnesota's forestland.

Jurisdictional responsibilities for forestland administration can be divided into four major ownership classes: federal, state, county and private (including Native American). The federal lands are managed under federal statutes for the management agencies and protected area systems.

By Minnesota statute, the Division of Forestry is the main agency responsible for managing state-owned lands 'according to the principles of multiple use and sustained yield' and pursuing 'the sustainable management, use and protection of the state's forest resources to achieve the state's economic, environmental and social goals'. Operational forest resources plans are developed with input from other divisions of the state Department of Natural Resources and the public through a formal planning process.

In most counties, land commissioners and their professional staffs are responsible for land management, which is guided by land management plans developed by each county. In 1979, the Minnesota Legislature enacted payment in lieu of taxes legislation to encourage retention of tax-forfeited land and to provide compensation to local taxing districts for loss in tax base as a result of this retention. Part of these payments is dedicated to intensifying forest and other natural resource management on these county-administered lands (Minnesota Association of County Land Commissioners, 2000).

Individuals, including farmers, own about 75 per cent of Minnesota's privately owned land (USDA Forest Service, 2002a). The remaining 25 per cent is fairly evenly divided between forest industries, other private corporations and Native American tribes. The federal Bureau of Indian Affairs and the Native American owners cooperatively manage forestland on these reservations. Forestland management activities vary widely on private lands. Forest industry owners generally manage their lands for intensive timber production, while other private owners manage their lands for a variety of purposes, including wildlife, outdoor recreation, aesthetics and timber production. Numerous federal and state programmes encourage private landowners to manage their forestlands in a sustainable way.

From the outset, stakeholder involvement has been an integral part of Minnesota's approach to sustainable forest resource management and protection. In 1989, a petition by state citizens concerned about the environmental consequences of expanding timber harvesting led to a study, completed between 1990 and 1994, that recommended dozens of tactical prescriptions for modifying land management practices, as well as numerous strategic policy options (Kilgore et al, 1996). A ten-member citizen advisory committee advised the Minnesota Environmental Quality Board in developing a state-funded generic environmental impact statement. In 1994, a 25-member group met for seven months to develop forest policy recommendations that could serve as the basis for legislation.

In 1995, the Minnesota Sustainable Forest Resources Act was enacted, and the Minnesota Legislature appropriated more than US$1.7 million to begin its implementation. The law authorized the appointment of a forest resources council charged with advising the governor and public forestry agencies on forest policies and practices, developing and implementing site- and landscape-level programmes across all ownerships, and coordinating research, information management and professional and public education (Kilgore et al, 1996).

During its first three years, the Minnesota Forest Resources Council and its many partners developed comprehensive voluntary timber harvesting and forest management guidelines. These guidelines are used by both public and private forest landowners, and over 85 per cent of the volume of timber harvested is by loggers who have received guideline training. Monitoring conducted since 1999 has served as the basis for comprehensive updating of the guidelines that will be completed in 2005.

Since 1998, landscape plans have been completed for four of six major forested Minnesota landscapes. A fifth plan is nearing completion, and a plan for the sixth region is under way. These plans provide a strategic context for coordinated use and management of forests across all public and private ownerships.

The Forest Resources Council faces some challenges, however, including funding for research, information management and education. Moreover, the environmental community has charged that actions to minimize environmental impacts of timber harvesting are inadequate and too slow, and logging businesses complain that implementing the guidelines is too expensive. Although some 525,000ha of forestland is third-party certified, a significant challenge remains to certify remaining state, county and private lands to meet the anticipated demand for certified forest products. Finally, there is the challenge of addressing the impacts of urban growth and development, outdoor recreational use, global climate change and continued timber harvesting and forest management.

Alabama

Alabama uses educational and technical assistance to promote sustainable forest management. This decentralized system relies upon voluntary adherence to best management practices for resource management and operations.

Alabama's forests have always been extremely important to its economy. Alabama has the second-largest commercial forest in the nation, with 71 per cent of the state (9.3 million hectares) covered in forests. Non-industrial private landowners own approximately 78 per cent of this commercial forestland. Forestry accounts for an annual payroll of over US$2 billion.

The federal government owns and manages four national forests encompassing 270,000ha, as well as several national parks. The state is responsible for managing close to 24,300ha of forestland and several state parks. State law to protect and enhance all forests of Alabama requires the state forestry agency to provide assistance to private landowners. Given the large proportion of forests owned by the private sector, these landowners have been critical for the success of sustainable forest management in the state.

All forest landowners must meet regulatory requirements of the federal Clean Water Act and Endangered Species Act. Furthermore, Alabama has formulated voluntary best management practices for forestry that also protect and improve water quality to meet federal guidelines, with recommendations related to zoning, harvesting, roads, crossings, wetlands protection and reforestation. Leaders in the forestry sector believe that Alabama's best management practices are critical components of sustainable forestry. Monitoring has shown excellent compliance through voluntary adherence to the recommended actions.

Besides certification schemes, several programmes have been developed to assist landowners – through both technical information and financial assistance – to ensure the sustainability of Alabama's forest resources. There are also associations and organizations that help and encourage landowners. These groups are composed of resource professionals and other landowners. They will continue to have a great impact on Alabama's forest sustainability by helping to address one of the major challenges facing Alabama: the increasing number of forest landowners associated with small tract size. Through these associations and organizations, Alabama's resource professionals reach more landowners to help them meet their goals and objectives.

A forestry programme called TREASURE (Timber, Recreation, Environment, Aesthetics from a Sustained Usable Resource) is a voluntary programme that promotes sound and sustainable, multiple-use forest management. Created in 1974, the programme encourages landowners to use their forests wisely in order to meet their own needs while at the same time protecting and enhancing the environment. It promotes a sound management ethic through two avenues: education and recognition. Education is provided through information and on-the-ground technical assistance by resource professionals (foresters and biologists). The landowner is guided and encouraged to work the land into a well-managed forest according to his or her objectives. Recognition occurs through the TREASURE Forest Award, which not only recognizes the landowner's hard work, but also encourages fellow landowners to do the same. Currently, Alabama has 1924 TREASURE forest landowners covering 735,000ha. There is a TREASURE Forest Landowners Association chapter in almost every county. In more rural counties, these groups are prominent in the community. Thousands of new landowners in the programme are working towards becoming a certified TREASURE forest landowner.

Conclusion

There is strong interest and concern for forest sustainability at all levels in the US. Several states have special initiatives, some with strong leadership by the governor, to ensure that all three pillars of sustainability are considered in forest management. The challenge is to provide not only long-term but also short-term human welfare, as well as equity among communities, both of place and interest, now and in the future.

This challenge will always be with us. Crucial to the ongoing solution will be an encouraging and enabling legal framework – one that recognizes the importance of stakeholder involvement at all levels, that enables them to be effectively involved in forest resource decisions, and that officially recognizes and provides credibility and accountability to the enabling process which is adopted and used.

Note

Gerald A. Rose would like to thank Catherine Karr Colque, of the US Department of Agriculture (USDA) Forest Service International Programs, for her assistance in editing and improving this chapter.

References

Fretwell, H. L. (2001) *Public Lands: Is No Use Good Use?* Political Economy Research Center (PERC), Bozeman, MT

GAO (General Accounting Office) (1999) *Western National Forests: A Cohesive Strategy Is Needed to Address Catastrophic Wildfire Threats*. GAO/RCED-99-56, Washington, DC

Jaakko Pöyry Consulting Inc (1992) 'Public forestry organizations and policies: A background paper for a generic environmental impact statement on timber harvesting and forest management in Minnesota'. Paper prepared for Minnesota Environmental Quality Board, St Paul, MN

Kilgore, M. A., Ellefson, P. V. and Rose, G. A. (1996) 'Innovative forestry initiatives: Minnesota prepares for the future'. *Journal of Forestry*, vol 94, no 1, pp21–25

Lee, R. G. (1994) *Broken Trust, Broken Land: Freeing Ourselves from the War over the Environment*. Book Partners, Inc, Wilsonville, OR

MacCleery, D. W. (2002) *A General Overview of Policy and Institutional Frameworks Governing the Forest Sector in the US*. USDA Forest Service, Washington, DC

Minnesota Association of County Land Commissioners (2000) *County Forest Management 1980–2000: An Application of Stewardship*. Minnesota Association of County Land Commissioners, MN

National Council for Science and the Environment (2001) *Recommendations for Achieving Sustainable Communities: Science and Solutions*, Washington, DC

Sedjo, R. A. (2001) 'From multiple-use to all-for-me'. *The Environmental Forum*, vol 18, no 3, p66

Steen, H. K. (1976) *The US Forest Service: A History*. University of Washington Press, Seattle

US Bureau of the Census (1975) *Historical Statistics of the United States from Colonial Times to 1970, Bicentennial Edition, Part 1*. Washington, DC

US Bureau of the Census (1994) *Statistical Abstract of the United States, 1994*. Washington, DC

USDA (US Department of Agriculture) Forest Service (2001) *Forest Resources of the United States, 1997*. General Technical Report NC-219, North Central Research Station, St Paul, MN

USDA Forest Service (2002a) *Forest 2002 Inventory and Analysis (FIA)-2002*. FIA Cycle 12 Annual Inventory, www.ncrs2.fs.fed.us/4801/fiadb/index.htm

USDA Forest Service (2002b) *The Process Predicament: How Statutory, Regulatory and Administrative Factors Affect National Forest Management*. USDA Forest Service, Washington, DC, June

Part III

Community Voices

Chapter 15

Decentralized Natural Resources Management in the Chiredzi District of Zimbabwe: Voices from the Ground

Steven Hlambela and Witness Kozanayi

Introduction

State-level and other external actors often define the nature and extent of power assigned to lower-level actors in most decentralization interventions. Such interventions predictably fall short of thoroughgoing empowerment because state-level and other external actors tend to retain meaningful sets of power. In this chapter, empirical evidence is used to test the assumption that decentralization is likely to result in more complete empowerment if it is demand driven. Should states retire entirely from the business of local governance, and can the local go it alone? This chapter reveals that decentralization, even if it is demand driven, is still no panacea for the problem of empowering local communities. Even armed with a clear road map of how they can initiate decentralization processes at the local level, rural communities cannot go it alone. They still need assistance in one form or another from district-level organizations and other external stakeholders. It is therefore imperative that the district-level organizations be active facilitators of decentralization at the local level, while not hijacking the process from the locals.

Decentralized natural resources management in the Chiredzi district

The call to put people first in environment and development initiatives has been advocated for some time (Chambers, 1983). Although the merits of seeking to put people first are now more readily appreciated and understood, in practice the community empowerment rhetoric has seldom been matched by real action. Putting people first, by and large, remains of symbolic significance in rural development research. The following sections present a bottom-up, local stakeholder perspective on a decentralization process in which power was sought from below, rather than granted from above, with a focus on land-use planning and access to resources. They present a community representative's own experiences and interpretations of a community-driven decentralization initiative in a Zimbabwean social forest in the Chizvirizvi resettlement area, which lies some 40 kilometres (km) east of Zimbabwe's south-eastern lowveld town of Chiredzi. The area of Chizvirizvi is about 23,440 hectares (ha) in extent.

After almost 15 years of state-inspired land-use planning initiatives, the community in Chizvirizvi abandoned the existing system of consolidated villages, in which densely settled villages are clustered together and separated from distinct grazing and cropping areas. The government pioneered this system of settlement in the post-independence era (after 1980) as part of its reconstruction and development policies. It was thought that such a zoning plan would help to alleviate population pressure in the surrounding communal areas while easing access to centralized services and amenities. The government failed, however, to deal with overcrowding. The need to decongest the consolidated villages saw communities implementing a new settlement plan based on 'private' and more spacious plots. Consequently, communities have been putting demands on the district-level government, requesting formal title for the plots and resisting payment of taxes until the council delivers on essential services. Communities have also demanded authority over local wildlife resources, now vested in the rural district council. Communities do not want the district councils to be responsible for the management of wildlife (as is the case with all wildlife rich areas). Instead, they want the community to be granted authority over the wildlife. In fact, the community of Chizvirizvi pursued this issue further and was granted appropriate authority (AA) over wildlife by the minister of environment and tourism in August of 2004. They become the first community (we know of) in the whole country to be conferred with AA status over wildlife.

Information on the community-driven resettlement plan comes from community representatives, district-level officials and other stakeholders, including a local commercial concern – the Malilangwe Conservation Trust. The trust provided funding support for the implementation of the community plan. The interpretation of the extent to which the new bottom-up plan (as well as the abandoned top-down plan) was seen to empower the local community is that of the senior author, who as the community leader introduced the new vision of settlement. We first provide a brief background on Zimbabwe's forests and the manner in which decentralization policies are unfolding.

Forest context and decentralization experiences

Zimbabwean forests fall into what is commonly referred to as the dry forest zone, which is dominated by *miombo* woodlands consisting primarily of the genera *Brachystegia*, *Isoberlinia* and *Julbernadia*. The forests have three major categories of ownership. In state forests, the state forestry agency, the Forestry Commission, practises 'exclusion management' – that is, no one is allowed to pass through such forests, let alone utilize their resources. Occasionally the government issues concessions to big private companies to log timber from the forests.

Woodlands on small-scale commercial farms and commercial farming areas are considered private property, dating back to 1930, when the colonial government set aside commercial farms (generally located in high rainfall areas with fertile soils for the whites) and communal areas (often dry, with infertile soils) for the majority African population. The present government inherited and perpetuated the old pattern of landownership until early 2000, when it earnestly started to redress the imbalances. State regulation of woodlands in the private domain was limited to controls against clear-felling for commercial purposes. In practice, however, such restrictions are largely un-enforced because of resource constraints.

The third category comprises woodlands in communal areas, including resettlement areas. The resettlement areas are a product of the post-independence government policy to acquire commercial farms for landless communal farmers. Tenants in resettlement areas receive government permits that entitle them to use the land and its resources; but in the case of Chizvirizvi, no such permits have yet been issued. Communal forests have subcategories of use and control. Trees and woodlands at homesteads and in crop fields fall under a de facto traditional freehold category, beyond which there are grazing and woodland commons. Use of grazing areas and woodlands is governed mostly by informal community rules, which may differ from place to place. Government departments – the Forestry Commission and the Department of Natural Resources – also enforce government policy on how woodlands may be used. For example, as a conservation measure, the government forbids cultivation within 30 metres (m) of a riverbank in any area.

Decentralization in Zimbabwe's forest sector has varied with the tenurial status of the land. In protected forest zones, collaborative resources management regimes have only recently been introduced. In this variant of decentralization, communities residing at the margin of state forests have access to some products from the forests. But access and use rights do not cover high-value resources (Wily, 1995), leading to resentment. Moreover, the significance of the forest margin zone is minor compared with other woodland types in the country and their importance for livelihoods. Collaborative resources management arrangements are thus restricted to the few communities at the immediate borders of the state forests.

Decentralization in the communal domain has tended to follow the contour of broader natural resource governance settings. Some laws with decentralizing orientations have actually recentralized power at the rural district council level. The Communal Lands Act of 1982 makes the councils *de jure* land authorities with powers to allocate land in communal areas that fall under their jurisdiction. The Communal Lands Forest Produce Act of 1987 vests control over indigenous woodlands in communal areas in such councils, as well. Through such legal entitlement, councils with commercially valuable timber species are given enhanced access at the expense of communities in communal areas. The Parks and Wildlife Act of 1975 bestows on the councils 'appropriate authority status', which vests custodianship of wildlife resources in councils where the wildlife is found. Through such authority, councils in wildlife-rich areas have been able to control and manage wildlife and benefits accruing from it at the expense of local communities. Communal Areas Management Programme for Indigenous Resources (CAMPFIRE) guidelines specify that producer communities should be entitled to 50 per cent of revenue collected from wildlife and other natural resources (Murombedzi, 1999; Dzingirai, 2003). Thus, the overall effect of decentralization in the communal domain has largely resulted in recentralization of power at the district level (Murombedzi, 1999; Mandondo, 2000). No law appears to decentralize powers to other levels of social organization, particularly those that are closer to the citizens.

In principle, there should be nothing fundamentally wrong with decentralization that focuses powers at the district level. The district is, in fact, the lowest unit of a local governance framework that encompasses popularly elected grassroots structures, including village development committees and ward development committees. These two structures were intended to spearhead development at the local level; but evidence suggests that they are used as political functionaries for the ruling party. The village development committee is presided over by an elected chairperson and represents 100 households, on average (Roe, 1995). Planning is supposed to begin at this level, and the plans of usually six villages are then consolidated into a ward development committee. This committee's elected councillor then becomes the ward's representative at the district level – the rural district development committee, which is dominated by bureaucrats and technocrats. In championing community rights at the district level, the councillors, mostly peasants, have to engage with those who have superior literacy levels and negotiating skills.

Overall, decentralization has tended to be championed by the government and other external actors, but has largely remained a form of conditional empowerment that does not lead to complete empowerment. The Chizvirizvi case marks an interesting departure from the norm in that it was community driven (compare the case of Scotland in Chapter 12).

Tragedy of the Chizvirizvi commons

Chizvirizvi adjoins a private wildlife conservancy and a national park, the Malilangwe Conservancy and Gonarezhou National Park, respectively. Decongestion and reconstruction and development policies in Chizvirizvi involved the implementation of a government-driven resettlement scheme. The government purchased land from adjoining commercial farms and created consolidated villages based on a system of land-use planning that divided landscapes into clustered villages, grazing areas and cropland. Each household was allocated 5ha of arable land, and grazing was communal. It was hoped that the centralized settlements would enhance peasants' access to water, electricity, road networks, schools, clinics, beer halls and grinding mills.

Ten villages, each presided over by a village chairperson, were established, all under the control of a government-paid resettlement officer. In total there were 283 households, mostly from the Shangaan tribe, which had been displaced earlier from the same area. The Shangaan group had maintained its culture as a tribe, but mingled with other tribes through intermarriage. Under this set-up, each household was allocated 5ha of arable land, and the remaining land (about 22,000ha) was designated as communal areas where livestock could be grazed and woodland resources such as firewood and construction poles could be obtained. The crop fields were situated in one area, away from the village compounds. Because all of the area outside the arable fields was communally owned, farmers from neighbouring rural communities could also use grazing and other forest resources from Chizvirizvi. Without a mechanism to monitor resource use, massive destruction of forest resources in the communal woodlands ensued.

Modest progress in providing services and basic infrastructure within the consolidated villages was undermined by the failure of the plan to address overcrowding. Social ills worsened: deprivation of individual and family freedom and autonomy, misunderstandings and fights, jealousy, thefts, suspicions of witchcraft, and adultery. The 5ha of arable land allocated per household was not sufficient even for the sustenance of the immediate family (some families are polygamous). Moreover, annual rainfall in this dry region is too erratic and unreliable to support rain-fed agriculture.

Local farmers therefore surreptitiously increased their crop fields beyond their allocations by encroaching on the communal woodlands. Social pressure thus became one of the major drivers that inspired the search for alternatives and galvanized action within the villages. Additional concerns included perceptions of rampant degradation of woodlands in areas surrounding the consolidated villages. This wanton destruction of woodland resources was carried out both by the locals and by people from surrounding communities. There was no monitoring of the pattern of use of woodland resources in the area: village heads had not been given the mandate to oversee the way in which local forests were used. A five-member committee of scheme residents then became the vehicle through which such concerns could be articulated.

The idea for the committee and for self-contained plots was advanced by this chapter's principal author (Hlambela), the district chair of the Zimbabwe Farmers Union, a powerful national organization that represents and champions the rights of all black farmers in the country. Because of his position in the union, the author has been able to interact with people at higher levels and in different fora. He has also received training in a number of development-related areas. At first, he proposed the idea of self-contained plots to a few influential people in the community, who readily accepted the idea. Later, together with other village elders, the author consulted the general community, which was also agreeable to the plan. The author was elected the director of the committee, who was to help the community achieve its vision. The local farmers subsequently elected four other members to this committee. Each of the ten villages nominated a representative to the Chizvirizvi Development Committee to represent their village.

A NEW DEMAND-DRIVEN MODEL

Community concern about overcrowding and environmental degradation in Chizvirizvi peaked around 1987. Such concerns marked the beginning of a community-led and demand-driven vision of land-use planning and natural resources conservation based on self-contained lots as opposed to overcrowded cluster villages.

Through the Chizvirizvi Development Committee, the community gained the support of the Department of Natural Resources for a more dispersed resettlement scheme in which the holders of individual plots would be ultimately responsible for most resources within their plots. Under the proposed model, plot owners would have exclusionary rights over their plots and could thus deter the wanton destruction of resources then occurring in the communally owned woodlands.

Most people were supportive of this proposal; but a few farmers were sceptical, especially since the community had to negotiate with high-ranking government officials. These farmers were especially doubtful of the ability of local representatives to articulate local interests at high levels. Support was then obtained from the Ministry of Lands and Agriculture in 1989. The support of now two government agencies inspired residents to put in place temporary mechanisms for apprehending violators and protecting their resources. Such measures included collective monitoring and policing of resource use.

Meanwhile, the Chizvirizvi Development Committee enlisted the support of the local chief and the chairpersons of the ten villages in question. This led to local-level endorsement, with signatures of endorsement. Plans were subsequently formally submitted to the government through the provincial Lands Office. Although the community received a favourable response from the government in 1995, the government did not have funds to support implementation. Through the author's links with the Zimbabwe Farmers Union and as chair of the Chizvirizvi Development Committee, the author was able to meet with the minister of agriculture to open up avenues for funding. Although no financial support could be obtained from the ministry, the Chizvirizvi Development Committee was able to secure ministerial endorsement that enhanced access to alternative sources, including donor support. On the advice of the Chiredzi District Administration, the community's donor outreach strategy emphasized building lasting partnerships with potential local donors. The strategy was meant mainly to avoid over-dependence and the time-bound nature of externally funded projects. Thus, building strategic partnerships was crucial as the community forged ahead with its vision of decentralized land-use planning and conservation of natural resources.

Building tactical partnerships

A meeting, organized by the Malilangwe Trust, brought together 49 community representatives and 15 visitors, mostly from the University of Zimbabwe but also from Zambia, Malawi, South Africa, Botswana and the US, to discuss 'Group versus Individual Settlement'. At that meeting, Malilangwe Trust committed Z$100,000 (US$2632) for surveys, mapping and demarcation of plots. However, not all the potential partners were willing to make pledges. For instance, a request for technical support in surveying and demarcating plots was met with derision from the local office of the national Agricultural Extension and Research Services, whose officers pointed out that they implemented plans only on instructions from above – not on instructions from below. Consequently, the community approached the provincial level, and a survey team was assigned to the area in June 1999. The community had to consider whether it was prudent to permanently sideline the Agricultural Extension and Research Services officers or bring them on board. The latter option prevailed as it was realized that cross-scale collaboration and diffusion of skills was essential for sustainability and continuity. The local councillor was also brought into the process on the basis of this consideration.

Overcoming inertia

Next, a planning meeting was held at Malilangwe, with community representatives, support teams from the local and provincial levels and the district administration. Of the total financial package received from the Malilangwe Trust, the community set aside Z$25,000 (US$658) for the survey and planning team. The money was handed over to the district administration to disburse to the team in stages as agreed tasks were completed. It was recognized that the survey team would need some means of transport. The Malilangwe Trust repaired a broken-down government Landrover at a cost of Z$12,000 (US$316) and made a cash donation of Z$4000 (US$105) for fuel.

However, no survey work ensued. On investigation, the community was shocked to learn that none of the resources had actually been given to the team. Speculation was that the money was invested in some financial scheme, with the proceeds being used for self-benefit. The Landrover was found, overturned and under a bridge, where it had been abandoned after being used for unrelated purposes. To ensure progress and some form of accountability, the community decided to return the funds and resources to the donor for allocation to the survey team against claims to be submitted by the team for tasks performed.

The survey and demarcation work started in late 1999. Further logistical support, including food for the survey team, was required. The community committed itself to providing mealie meal and labour, and the Malilangwe Trust pledged 4 kilograms of beef per day. After the survey and demarcation work were completed, occupation of the plots was delayed, apparently as several actors, including the district administrator and local leaders, purportedly vied for credit, recognition and influence. During the delay, unfortunately some demarcation pegs were destroyed, thus requiring further investments in repair and maintenance. The allocation of plots finally got under way in March 2000. Other than community members, the event was witnessed by the district administrator, the local member of parliament and two representatives from the president of Zimbabwe's office.

Allocation process

A total of 294 plots, each measuring about 85ha, were to be allocated to households from the ten villages of the original resettlement scheme. Allocation was done through a raffle in which people randomly picked cards whose numbers corresponded to particular plots. This method of allocation had been used earlier by a local farmers' group to share fairly among themselves cattle of varying sizes and ages.

First preference for the new plots was given to people formerly residing in the government-initiated consolidated villages. Some of these people were initially reluctant to move because they had invested in infrastructure at their homesteads; but they joined the last-minute rush for the plots, when plot ownership was opened to anyone. The rush grew in volume when good crop harvests were secured by the pioneer group of settlers, and power plays emerged as people asserted claims to the land. A group of teachers at one of the local schools was one strong constituency that had been left out of the initial allocation process. They clandestinely instigated the local village worker to mobilize people in two villages to revolt against the scheme, purportedly because they had been allocated infertile plots. In the hope of limiting the ensuing conflict, the district administrator unilaterally took over control of the unassigned plots. The custodianship did not last long; but neither did it dampen the conflict. Realizing the futility of intervention, the administrator re-vested control in the Chizvirizvi Development Committee. To break the impasse, the Malilangwe Trust organized a multi-stakeholder meeting to which the various adversaries were invited. At that meeting most stakeholders' concerns were amicably addressed, and some of the teachers received plots.

A group of 16 liberation war veterans also felt that they had been left out of the entitlement process. The chair of the land committee decided to enlist the support of the chief in deciding how to handle the issue, and they agreed that the land should be allocated to them as soon as possible. But there were only seven plots for 16 veterans. With the concurrence of the veterans, it was decided that the plots would be allocated through an elimination raffle. Each veteran would randomly pick cards marked either *yes* (a plot) or *no* (no plot). Although the raffle had the initial support of all concerned, the losing veterans did not honour the verdict, and most of them grabbed plots allocated to other people. One of the war veterans annexed the plot that had been assigned to the district administrator and promptly started development.

Some conflicts had legal implications. One woman annexed plot 83, which had earlier been allocated to the youth to carry out their own cash-generating projects, and enlisted the support of lawyers to resist her eviction. Eventually, four members of the land allocation committee were sued when they remained steadfast in endorsing the youths' claim.

Intra-family conflicts over landownership have started to surface in Chizvirizvi. Such wrangles usually emerge after the death of the plot owner when the question of inheritance arises. In one case, a couple divorced and the wife chased the husband away from the plot. The plot is registered in the wife's name because at the time the plots were allocated, only people who were not formally employed were eligible to apply, and the man in question was employed and thus ineligible.

Farmers from neighbouring communal areas are also claiming access rights (especially to graze livestock and collect firewood) from the self-contained plots. They argue that they had enjoyed such user rights when the farm was privately owned and unfairly lost them when the resettlement scheme was implemented. They also allege that some farmers from drier and rocky sections of Chizvirizvi cross over to the communal areas to get resources such as water and fertile fields, which gives the farmers from neighbouring villages the right to access resources from Chizvirizvi, in the spirit of 'reciprocity'. The situation is complicated further because many plot owners have ties with farmers from the neighbouring communal area: both groups of people are of the Shangaan tribe. It is therefore very difficult from a social point of view for the plot owners to deny 'brothers' access to woodland resources.

Some people feel that the Chizvirizvi Development Committee overstepped its role. And neighbouring rural communities view the programme of self-contained plots as an 'un-African' attempt by the plot owners to deny those communities access to God-given woodland resources. In any case, they feel they have longstanding ties of kinship that date back to their place of origin, South Africa, and should therefore share the little they have as brothers. Several contested plot ownership cases have been precipitated by the death of the male plot owner. The community has proposed that where the plot is registered in the husband's name, the spouse automatically inherits the land. However, some plot owners fear that under such a rule, a widow might remarry an outsider who would then take over the plot, thereby denying the deceased's relatives an opportunity to inherit the land.

All in all, the locally initiated decentralization planning has not been a smooth process. As a community several challenges were encountered, from inception of the idea of self-contained plots through to the final allocation of plots. The community also learned lessons from the locally initiated land-use planning model of decentralization. The section that follows highlights those, as well as some of the perceived benefits of decentralization.

Lessons learned

Decentralization in one area can provide a lever by which community interests can be mobilized in a wider range of arenas. Having empowered ourselves to effect changes in land-use planning, we have begun demanding parity between taxes and services. We also seek an opportunity to assert our interest in wildlife and appropriate authority over this resource, as well. Decentralization can also result in the creation of new local-level institutions; but continuous creation of institutions results in the overcrowding of the local institutional landscape. We also find that there is scope for engaging other external actors in resolving the various conflicts.

Land-use planning

Chizvirizvi presents a very interesting case in which a community opted to move from a top-down, government-sponsored system of settlement in favour of a bottom-up, community-driven model of settlement. The experience should serve as a poignant reminder to government and other external actors that no matter how well intentioned, top-down plans that do not accord with people's priorities and aspirations are bound to fail. Among other things, the consolidated villages were meant to enhance access to services and amenities and to bring some order to the pattern of rural settlement. These state-envisaged benefits, however, were at variance with the realities of reorganized settlement. Despite marginally improving access to services, consolidated villages, in fact, aggravated a range of social ills. The dense settlement also led to rampant destruction of woodlands and resources around the village sites. We valued the autonomy and independence associated with self-contained plots more than the order, neatness and access to services that the government sought to enhance through consolidated villages.

The lesson is that for any initiative to endure, it must be relevant, and to be relevant, an initiative must be based on people's needs and priorities. It also shows that there is a higher likelihood of success if the changes are bottom up and community driven rather than top down and government driven (compare Chapter 12 on the Scottish experience). We have realized that our collective will and resolve are assets through which we can empower ourselves, and we have started asserting our interests over a wide range of areas.

Taxation

We, the farmers in Chizvirizvi, are refusing to pay taxes to the rural district council because there is no parity between the tax collected and services provided to the community. In particular, we are irked that even as the council seeks to collect taxes from us, they are failing, if not refusing, to help us obtain title deeds for our plots. Title will ensure a high degree of security, giving farmers a legal document to exercise exclusionary rights to resources in their plots. Title deeds will also be used as collateral to secure loans from banks.

We have realized that if we are united at the local level, we have power to force institutions at the district level to be accountable to us. Through local leadership structures and the ward councillor, we can try to influence how our taxes are used. But this is possible only if we feel empowered to engage district-level stakeholders. Unfortunately, the district-level stakeholders seem to be reluctant to grant our requests, perhaps even in some cases acting in a retaliatory manner. The council is allegedly blocking potential donors from assisting the Chizvirizvi community because they consider us 'self-sufficient'. In addition to demanding parity between taxes and service delivery, decentralization has also accorded the community an opportunity to assert our interest in wildlife, particularly with regard to conferring appropriate authority directly upon us.

Wildlife management

Our present thrust is to ensure that entrustments from decentralization initiatives are conferred on other levels of social organization, rather than on districts alone. We want to ensure that entrustments, which are vested in bodies that are far from the citizens, are subject to debate and tenurial and use pressure. The owners of resources should also be the users and managers of such resources. Decentralizations in which roles and responsibilities are left hanging at a variety of levels do not result in empowerment.

Currently, only Malilangwe Conservancy and the rural district council reap the benefits from wildlife, while we pay the costs through crop losses. More often than not we are not even compensated for the damage. We are therefore requesting that appropriate authority over wildlife be conferred on us so that we can decide how benefits from wildlife can be used. When authority, such as authority over wildlife, devolves to local communities, it is important that appropriate local institutions be identified for the specific task.

Local institutions

Decentralization is usually accompanied by the creation of new institutions to spearhead the decentralization initiative. Creation of several institutions usually results in conflicts. In the case of Chizvirizvi, the government came up with new positions, such as the resettlement officer and village chairpersons, and we, the local people, came up with the Chizvirizvi Development Committee to guide the community in its attempt to initiate and implement a locally derived land-use planning exercise. The ruling party has also superimposed its own party structures in the area to safeguard and promote party policy. The creation of several institutions at the local level has resulted in feuds, especially between the village heads and *kraalheads* (traditional leaders of hamlets), who seem to be duplicating each other's role. The new Wildlife and Development Committee chaired by the ward councillor has literally usurped the roles of the Chizvirizvi Development Committee, which we created to spearhead the implementation of the new scheme.

Because of the prevailing political climate in the country, most local institutions, even those that would traditionally have a development mandate, have been politicized and are conduits to spread party propaganda rather than agents of development. When power is devolved to local communities, it is therefore important that new institutions with a development mandate are formed to spearhead development. Alternatively, existing structures should be given the resources to handle their new roles.

Conflicts

We readily acknowledge that our experience, though community driven, was not smooth sailing. We experienced a variety of conflicts at the community level as well as district level. We are aware that the conflicts mostly arise from jostling for benefits, influence, recognition and legitimacy. At the local level, most conflicts involved contests over prime land and resources during the allocation of the plots. We also experience routine boundary conflicts among plot holders, as well as between plot holders and residents of the adjacent communal area.

Contentious issues that appeared to put us in conflict with the district level include what we perceive as the failure and reluctance of officials at this level to formalize entitlements to our plots, as well as to vest us with direct authority over wildlife resources in our area. We realize, however, that conflicts are an inherent feature of most communities, and that they are likely to emerge whether decentralization is top down or locally driven. We believe that there is a need to put in place mechanisms to amicably deal with such conflicts. Suppressing conflicts merely defers the problem. We see conflicts at the community level as being best solved through approaches that seek to clearly define the roles of the different institutions. We believe that politically inclined institutions are not the best champions of developmental agendas.

Alliances with external organizations

The Chizvirizvi initiative was essentially community driven from inception to implementation. But to what extent can community members go it alone in planning and implementing community-driven decentralization initiatives? Although it was our vision, and although we mobilized residents to gain wider acceptance for our plan and organized local labour for the pegging of plots, we could not have done it without the support of external stakeholders. Training in leadership and project management and participation in workshops can help local leaders, as it did the principal author. We received financial support from the Malilangwe Trust, which also helped to mediate conflicts within the community, as well as between us and other stakeholders. We benefited from technical support from relevant government departments, and legitimating support came from a range of politicians, including a vice president, the minister of agriculture and provincial and district administrators. In initiating decentralization and other rural development initiatives, there are many roles that can be taken up by external actors; but the agenda should be set by the community members.

Conclusion

Decentralization initiatives that are demand driven seem to be enduring at the local level. First, the community should have a common vision of what its members want to achieve. The community should be led by visionaries who may not necessarily be part of existing leadership structures. Even though decentralization initiatives are started and driven by the rural communities, the communities should also be assisted by external organizations, including the district-level structures from which the communities are trying to emancipate themselves. However, external actors should only be facilitators and not highjack the whole decentralization initiative.

Decentralization can also spawn conflicts at the local level as new responsibilities and opportunities arise with the devolving of new powers to the community. Conflict resolution mechanisms should therefore be put in place to resolve problems as they arise, and the roles of institutions should be clearly defined and explained to all the institutions.

Note

This chapter is a result of research work funded by the Center for International Forestry Research (CIFOR). However, the views expressed are those of the authors. The authors are very grateful to Alois Mandondo for tirelessly commenting on the different versions of the chapter. We also want to thank Godwin Kowero for his insights at every stage of the fieldwork and writing. Anonymous reviewers who commented on the chapter and gave us a lot of thoughtful insights also deserve special mention. Last but not least, we thank the people of Chizvirizvi, who freely gave us vivid accounts of their experiences and views of decentralization processes in their area.

References

Chambers, R. (1983) *Rural Development: Putting the Last First.* Longman, London

Dzingirai, V. (2003) 'The new scramble for the African countryside'. *Development and Change*, vol 34, no 2, pp243–263

Emery, M. R. (1995) 'More than just politics of decentralization: Local government reform, district development and public administration in Zimbabwe'. *World Development*, vol 23, no 5, pp833–845

Mandondo, A. (2000) *Situating Zimbabwe's Natural Resources Governance System in History.* Occasional Paper No 32, CIFOR, Bogor, Indonesia

Murombedzi, C. J. (1999) 'Policy arena: Devolution and stewardship in Zimbabwe's CAMPFIRE Programme'. *Journal of International Development,* vol 11, pp287–293

Wily, L. (1995) 'The emergence of joint forest management in Tanzania'. Villagers and government briefing paper prepared on behalf of the Regional Forestry Programme and Shingida Rural District Council, Tanzania

Chapter 16

Decentralizing Protected Area Management at Mount Kitanglad

Adolino L. Saway Alyas Datu Makapukaw
and Felix S. Mirasol Jr

Introduction

Management of the Mount Kitanglad protected area in the Philippines was shifted from a central government authority to a board whose membership includes representatives of local stakeholders. Among the problems encountered are financial, language and cultural barriers to full participation by the indigenous representatives. Moreover, rivalries have arisen among tribal leaders. Nevertheless, this form of decentralization has reduced pressure on the forest and improved forest law enforcement. Community-based forest protection volunteers, the issuance of tenure instruments to qualified people and the accountability of tribal leaders for unsustainable natural resource extraction are the primary reasons for the success.

Mount Kitanglad protected area

Dominating the north-central portion of the Bukidnon Plateau, the Kitanglad Range covers a total land area of 47,270 hectares (ha) in the seven municipalities and one city of the province of Bukidnon, the Philippines. It plays a vital role in the socio-economic development of north and central Mindanao. Aside from being the homeland of the indigenous peoples of Bukidnon, it is also a major watershed providing water for power generation, irrigation and domestic use for Bukidnon and adjoining provinces.

The Mount Kitanglad protected area was chosen as one of the ten priority sites under the World Bank's Conservation of Priority Protected Areas project. On 9 November 2000, Mount Kitanglad finally became a full-fledged protected area when the Philippine Congress approved Republic Act 8978 – the Mount Kitanglad Range Protected Area Act of 2000.

Decentralization policies

NIPAS Law Republic Act 7586 of 1992

Decentralization of protected area management in the Philippines came only with the passage of Republic Act 7586 on 1 June 1992. Known as the National Integrated Protected Area (NIPAS) law, this legislation shifted protected area management from the Department of Environment and Natural Resources, a national government agency, to the Protected Area Management Board. This local, multi-sectoral body is chaired by the Department of Environment and Natural Resources with representation from the local government units, non-governmental organizations (NGOs) and civic organizations, people's organizations, indigenous peoples through their councils of elders and national government agencies.

The Protected Area Superintendent Office, which serves as the secretariat to the Protected Area Management Board, is also responsible for implementing all plans, policies and projects approved by the board.

1997 Indigenous Peoples Rights Act (Republic Act 8371)

The enactment of the Indigenous Peoples Rights Act (IPRA) in 1997 is another milestone in environment and natural resources management. It provides a solid policy direction that recognizes and promotes the rights of indigenous peoples, who are the primary occupants of the protected area. A certificate of ancestral domain title over ancestral lands is granted to genuine, qualified indigenous tribal communities. The title guarantees ownership over ancestral domains after the area has been identified and delineated in accordance with the law. The administration and management of certificates of ancestral domain title are based on traditional rights and practices. The traditional property regimes exercised by the indigenous people under their customary laws govern the relationships of all individuals within their communities.

To carry out the policies set forth in the Indigenous Peoples Rights Act, the National Commission on Indigenous Peoples was created to serve as the government agency primarily responsible for formulating and implementing policies, plans and programmes identified under this law.

1991 Local Government Code (Republic Act 7160)

Republic Act 7160, otherwise known as the Local Government Code of 1991, is the landmark legislation that introduced changes in the political process and transfer of power and authority from the central government to the lower local levels of government. The code expresses full adherence to decentralization as a matter of state policy and stipulates that local governments shall enjoy meaningful autonomy in order to attain their fullest development as self-reliant communities and make them more effective in attaining national goals. To implement this policy, the code declares that the state shall provide for a more responsive and accountable local government structure instituted through the system of decentralization. This means giving more power, authority, responsibilities and resources to local governments.

CONSEQUENCES OF DECENTRALIZATION POLICIES

As this enormous effort gets under way, the Protected Area Management Board, national government agencies, local government units and the indigenous communities are grappling with ways to proceed as successfully as possible. Various concerns affecting their respective programme implementation are described below.

Lack of collaboration

There seems to be difficulty in establishing a solid foundation for collaboration. Each management entity seeks to hold on to its own management regime apart from the others. Some tribal communities opposed to the Protected Area Management Board arrangement tend to ignore the mandate of the NIPAS law and to hold on to the IPRA. The National Commission on Indigenous Peoples, charged with implementing the IPRA, is faced with tremendous complaints from all parties, including the local government units. Since IPRA is a new law, the organizational requirements are not yet effectively in place and most of the needed resources are not yet available. This condition led to a struggle among the local government units, the Protected Area Management Board and some tribal leaders concerning whose policies should be implemented on the ground.

Overlapping management responsibilities

Developing a single vision is constrained by political and methodological hurdles. Management regimes in the context of the NIPAS law, IPRA and even local governments have long held limitations within their specific mandates and geographical jurisdictions. NIPAS concerns are bounded by their geographic scope. The IPRA recognizes that ancestral domain management has both geographic and cultural components. Likewise, the local government has its own geographic and administrative limitations. In most cases, these three management entities experience overlapping management responsibilities because protected areas are often claimed as ancestral domains, and some parts of the ancestral domain are definitely within municipal territories. The chain of responsibility and command has, to some degree, resulted in conflicts of interest and isolation.

The passage of the Mount Kitanglad Act institutionalized partnerships among the government agencies through the Protected Area Management Board. The constraining factor is that national government agencies have not harmonized their programmes, resulting in uncoordinated priorities. The board challenge is to ensure participation in the park decision-making process to gain support from various stakeholders. Initial results are encouraging; but the pace is slow as the organizations involved have different structures and need time to understand the essence of collaboration.

IMPLEMENTING DECENTRALIZATION

Major accomplishments

The Mount Kitanglad protected area is fortunate to have been named one of the ten priority sites under the World Bank-funded Conservation of Priority Protected Areas project. The project pilot-tested the decentralization provision of the NIPAS law for seven years, starting in 1994. The following are the major accomplishments that contributed to implementing the government decentralization policy:

- Designation of Mount Kitanglad as a protected area. In accordance with the NIPAS law, the Philippine congress set aside Mount Kitanglad as a permanent protected area through passage of Republic Act 8978. The law is a product of social legislation because it involved the participation of local stakeholders.
- Preparation and approval of the Mount Kitanglad Protected Area Plan.
- Creation and operation of the Protected Area Management Board.
- Issuance of tenure instruments to qualified tenured migrants and interested indigenous peoples.
- Establishment of the Protected Area Trust Fund. A Mount Kitanglad trust fund was established for the purposes of promoting sustained financing of the park. To date, at least 1 million pesos (US$20,000) has been generated from the visitors' entrance and land-use fees. The Protected Area Management Board will decide on the allocation of the fund, which is limited to park protection and community livelihood development activities.

Community participation: The council of elders

The Protected Area Management Board, with support from NGOs, has assisted tribal communities in identifying and establishing their councils of elders. The creation of councils of elders is mandated under existing laws and follows the inherent community structure. The process of identifying and affirming the respective chair of each council of elders started at the community level. The elders bring the concerns and interests of their indigenous communities to the Protected Area Management Board. Likewise, they serve as advisers and experts in formulating policies that are in harmony with their customary rights and practices. They are responsible for transmitting information to the indigenous communities who are not represented on the board.

The council of elders plays a major role in protected area management, particularly in planning, decision-making in resource access following their traditional practices, resolving boundary conflicts, and recognizing and respecting customary rights and practices. The elders identify and determine genuine indigenous knowledge systems and their differences from one village to the other. A ritual requirement that varies from one entry point to the other is an example of tribal differences that were considered in the formulation of Protected Area Management Board policy.

One of the challenges faced by the council of elders is the proliferation of tribal leaders and other educated members of the tribe who are seemingly working to control indigenous people's activities in the park. In order to remedy the situation, the chair of the council of elders was tasked to go around the park's seven municipalities and city to conduct cultural mapping and identify genuine indigenous leaders who could play a representative role in the Protected Area Management Board. The Protected Area Superintendent Office, the National Commission on Indigenous Peoples and the local government units assisted the chair in this activity.

After the World Bank project ended in 2002, funding for the council of elders operation ceased, except for limited activities attached to the Protected Area Management Board's annual work plan. Thus, it is difficult for the indigenous people's representatives to perform their tasks with the board and within their communities. The respective local government units have committed to support their work; but not all of them have the capability to deal with the government processes necessary before any assistance can be made available. There is also the added challenge of increasing the elders' negotiating and speaking capabilities so that they can hold their own during discussions with other board members. Service providers outside the government agencies are negotiating to pool their efforts and resources to address this concern.

Organization of the Kitanglad Guard Volunteers

The Kitanglad Guard Volunteers is the community-based park protection arm of the council of elders and the Protected Area Management Board. They serve as the tribal guards (*alimaong*) based on their tribal community structure. They become members only after they undergo a ritual performed by the elders and are deputized by the board.

In order to sustain the operation and provide support to the volunteers, the provincial government has allocated 1 million pesos every year to support their activities and provide incentives to performing members. Likewise, the provincial vice-governor covered their insurance benefits, and the municipal and city governments and some private companies have provided them with logistical support, such as raincoats, flashlights and other equipment.

One of the problems brought to the council of elders by the Kitanglad Guard Volunteers is that they are treated as rivals to the guards organized by the tribal leaders who are opposed to Protected Area Management Board operation. There is now greater pressure on the National Commission on Indigenous Peoples, the government agency tasked to implement the IPRA, to guide and mediate the conflicting interests of the natural resource protectors.

Local government units

After the completion of the World Bank project, local government units assumed responsibility for sharing the funding of activities in the approved management plan. A total of 6.7 million pesos (US$134,000) was allocated for 2002 and 2003 from the city, municipal and provincial governments. They also supported training and similar activities at the community level. The municipal and city chief executives and the provincial vice-governor personally attended Protected Area Management Board meetings to facilitate easy and immediate decision-making.

National government agencies

The Department of Environment and Natural Resources provides the overall supervision of the park through the Protected Area Management Board. The National Commission on Indigenous Peoples assists indigenous peoples' communities in asserting their rights and the exercise of traditional practices, and in the process of applying for ancestral domain title within the park. To date, at least seven applications have been received. The Department of Agriculture provides training, promotes sustainable upland farming systems and offers planting materials. The 28 *barangays* (villages) are currently the targets in this programme, with the establishment of one demonstration farm per *barangay*.

Private companies

The fees that private firms pay for resource use accrue to the park trust fund. The funds are used to sustain park operation and livelihood proposals of the local communities. Private companies likewise support the annual gathering of the tribes, the *Aldaw Ta Kitanglad* ('The Day of Kitanglad') Celebration.

ISSUES AND PROBLEMS OF DECENTRALIZATION

Experience

NIPAS law implementation in Mount Kitanglad Protected Area began in 1993. After almost ten years of implementation, we can describe the actual experiences from the affected communities as follows.

Misrepresentation

People living in the *barangay* proper and those close to politicians were chosen to represent the local communities in the management board and local government unit committees; but they do not truly advocate for them. This is perhaps one of the reasons why some of the decisions made for the indigenous peoples and local communities did not address their concerns or gain their support. In some cases these representatives did not earn the trust and confidence of the communities because of mishandling of funds and other similar acts. It is hoped that the cultural mapping initiated by the chair of the council of elders will address this issue by bringing in new representatives who have been chosen and fully endorsed by the concerned communities.

Limited funding support

Although the law mandates proper representation of communities on the Protected Area Management Board, most of the time their effective and meaningful participation is constrained by limited funding. This includes support for transportation, consultation, getting feedback and capability development. The board has taken up this issue with the local government unit chief executives, who are board members. Only 1 per cent of the local government units' internal revenue allotment is now set aside for the indigenous people. The challenge is to increase the local community representatives' capacity to access this allocation following the processes required by the local governments.

Insensitivity to the local culture

The local community representatives have found it difficult to actively participate in discussions because most members speak English at the meetings. The local dialect, or *binukid*, is seldom used, and most of the members and the Protected Area Management Board secretariat cannot comprehend it. Moreover, the board's manual of operations is not necessarily in harmony with the local culture. Tribal communities decide by consensus, whereas the board decides by majority vote. Most of the meetings are conducted in the *barangay* proper and seldom in the *sitio*, where most of the park occupants reside. If they do attend meetings, they do it for compliance because of their limited comprehension. They perceive Department of Environment and Natural Resources personnel as forest law enforcers who run after them for investigation; hence, they shy away from attending meetings called for by the agency. They usually abide by decisions even against their will as they do not want to be branded as rebels.

The Protected Area Management Board has recognized the problems and taken the following steps:

- Speaking *bisaya*, not English, during Protected Area Management Board meetings has become mandatory. The majority of the indigenous people's representatives speak *bisaya*.
- The number of indigenous people's representatives, chosen based on their customary practices, has increased.
- Community and/or cultural mapping will determine a local community's location and features.
- The services of the elders have been engaged for project monitoring and information dissemination.
- The tribal justice system is enforced before forestry laws.
- Cleansing rituals are performed in every activity undertaken with the community. This is to ensure indigenous people's trust and support.

Unequal opportunity for livelihood projects

Livelihood projects are usually awarded to the *barangay* community members and only in a very limited manner to the protected area residents. This has happened because some of the communities are not included in the park census and registration; they perceive the census as a way for them to be identified and to bring them to court. Furthermore, most of their area does not qualify for livelihood criteria, since it is very remote and monitoring is difficult. Those who are included encounter difficulty in withdrawing funds from the bank because they lack identification and their signatures vary.

Opportunity to increase knowledge and interaction

Indigenous people's representatives who are already working with the Protected Area Management Board have found that their participation is an opportunity to increase their understanding and view of the other side of development. Their interaction with the other park stakeholders has given them an opportunity to make new contacts and to educate partners on the importance of their customary practices. Gradually, they find their initiatives rewarding as most of the decisions pertaining to indigenous people's concerns have been favourably acted upon, including board resolutions and *barangay* ordinances. The field exposures and cross-visits with the other board members have increased their perspective in viewing developmental strategies in the context of partnership.

Decentralization of protected area management requires participation from the different sectors and groups. This venue provides an opportunity for stakeholders and members to meet tribal communities from the other side of the park. The annual celebration of the *Aldaw Ta Kitanglad* is a two-day gathering of almost 800 people who are stakeholders in the park. The activities are appropriate to the indigenous peoples' customs and traditions when they gather together in important occasions and give elders an opportunity to educate the younger generation and non-members of the tribe.

Major issues and problems

Inequitable sharing of national wealth

There is no equitable sharing of revenues derived from the national wealth. Only national and local government units have a share through their annual internal revenue allotment. Funds are not directly allocated to the indigenous people following their tribal structure, but are instead channelled to the *barangay* treasurer. Should there be funds from the local government units, they tend to be used for project implementation – seldom priorities for the indigenous people.

Confusion over legal provisions

The NIPAS law, IPRA and local government codes were written without grassroots consultation. This resulted in misunderstanding and conflict between some indigenous communities and government agencies. Unwritten (customary) laws are prone to abuse by some educated tribal leaders seeking to advance their personal interests. This leads to confusion among tribal leaders and other stakeholders of the park about when to apply customary law and the criminal justice system. In some instances, this is the source of conflict and, in fact, has created factions among them. The National Commission on Indigenous Peoples, which is supposed to guide the indigenous peoples in the smooth implementation of the laws, is preoccupied and can no longer address all of the demands. Some private companies benefiting from park water resources see opportunities to work with the community in order to ensure sustainability of their operation, provided it is reasonable and within their capacity. Others, however, use the 'free and prior informed consent' requirement to exploit the indigenous people, and there are no clear parameters on what the community can demand.

Taxation is a common source of confusion between the Protected Area Management Board and local government units: both have mandates to collect revenues in the same area. Another issue is who should manage the park after an ancestral domain title is awarded to the indigenous peoples. Will it be the board, as mandated under the Mount Kitanglad Act, or the council of elders, as provided under IPRA? The Department of Environment and Natural Resources and the National Commission on Indigenous Peoples have drafted a joint memorandum instituting mechanisms to harmonize the laws' perceived conflicting provisions; but implementation of the policy largely depends upon the capacity of the fieldworkers to appreciate and comprehend the real intent of the law.

Different interpretations of IPRA

The whole area of Mount Kitanglad Range was claimed and applied for by three tribes under a unified ancestral domain title in 1995. The application was met with apprehension, and conflicting interpretations of IPRA hindered the government from taking timely action. Some of the problems encountered include overlapping claims and opposition raised by other tribal communities. In frustration and, perhaps, having lost confidence in the government agencies, some indigenous leaders are no longer on speaking terms with the government. Some of the frustrated leaders have started organizing their own tribal guards and strengthened their campaign for better enforcement of the free and prior informed consent provisions. Likewise, they have openly manifested their intention to boycott participation in the protected area management until such time as their claim to title is awarded to them and other customary regulations are settled.

Worse, projects and activities of the local communities who are not supporting their cause are being blocked for implementation. Although this does not have much impact on the ground and affects only one of the eight local government units' political jurisdictions, this move will certainly divide the tribe. The Protected Area Management Board, after the opposing party put down its initiative and closed the lines of communication, decided to concentrate its effort on communities who support the government programmes. Likewise, the National Commission on Indigenous Peoples has also prioritized these communities. The board and other concerned groups are hopeful that, in time, all parties will understand and accept the policy direction and exercise the true intent of IPRA.

Results of decentralization

Forest protection and natural resources management

Decentralization has reduced pressure on the forest and its natural resources and effectively improved forest law enforcement. This change is due to the presence of the community-based forest protection volunteers and the issuance of tenure instruments to qualified peoples. Likewise, tribal leaders became responsible and directly answerable to the management board for any form of unsustainable natural resource extraction. Furthermore, with the participation of most stakeholders in the Protected Area Management Board, there is an increased awareness and a process that promotes checks and balances. Fire incidence has decreased, from 103 occurrences recorded in 1998 to only 12 in 2003, as has the area damaged by fire. Based on 21 biodiversity monitoring reports and focus group discussions among trained local monitors, the number of wildlife species has increased.

Livelihood assistance and revenue generation

Livelihood assistance is made available to the local communities as per the protected area management plan. The Protected Area Management Board has brought in local and foreign institutions to help implement socio-economic activities under the approved plan. Since decision-making has been passed down to the local stakeholders, this promotes more equitable distribution of benefits and effectively identifies the target beneficiaries, particularly those living within the buffer zone. Budgetary allocations from local government units have been made available to the park, revenues generated out of the park operation accrue to the park's trust fund, and linkages with funding institutions have improved.

Multi-sectoral participation and collaboration

Decentralization provides an opportunity for the local stakeholders to take part in all decision-making processes and governance of the protected area through the Protected Area Management Board. The creation of the board promotes an atmosphere of collaboration and offers a place where conflicts can be resolved. What is left unresolved to date is the issue of the unified ancestral domain claim. This is because the head claimant refuses, out of frustration, to bring the issue to the board for discussion, and some concerned offices have failed to understand the interest and intention of the claimant. An effort is now under way to open lines of communication to the majority of the indigenous leaders who are supportive of the government programmes and to identify the genuine tribal leaders of the park through cultural mapping. This effort has the support of the local government units, the Department of Environment and Natural Resources, the National Commission on Indigenous Peoples and the chair of the council of elders. Meaningful participation, awareness and partnership from the local communities and other stakeholders of the park are a continuing challenge to ensure the effective enforcement of policies and projects on the ground.

Tribal leaders

The Protected Area Management Board's recognition of the authority of tribal leaders has caused the unprecedented proliferation of such tribal leaders. In some instances, a leader is tempted to use his power and authority to entice people from outside to own land within his territory at a price. Genuine leaders, in some cases, are also struggling for power, and the usual loser is the uneducated over a schooled member of the community. Personal interest, greed, desire for power, and positioning to corner project assistance from funding institutions in the guise of customary rights and protection of indigenous interests are common. One challenge is to codify genuine customary practices so that these cannot be abused and used to exploit others.

Council of elders funding

The sad reality is that there are really no funds allocated to sustain the operation of the councils of elders despite their very important role of promoting peace and order, planning, education and decision-making. They are very dependent upon the local government units and other government agencies for support.

Local government's role

The chief executive of each of the local government units down to the *barangay* level is a member of the management board. This arrangement gave them an opportunity to get firsthand information and to become part of the planning and decision-making process in park management. Through this the local government units were able to consider protected area plans in their respective annual investment plans and have the resources allocated to implement the priority programmes of the park. Since 2002, at least 6.7 million pesos (US$138,000) have been allocated to the park from the local governments. Currently, all the municipalities and 28 *barangays* have approved their respective forest use plans and the watershed management planning framework. On the ground, the tribal leaders and *barangay* captains are accredited as Protected Area Management Board point persons; thus, forest management is within the reach of the people. They can decide on matters of boundary conflict, resource access for domestic use, visitors' entry, disaster preparedness and recommend volunteers for deputation. *Barangay* ordinances are being passed to complement board policy if necessary.

Lessons learned

- Park management can be implemented successfully by changing the locus of decision-making from national agencies to local levels, emphasizing multi-sectoral participation.
- Decentralization is not primarily the domain of the local government. The main essence of decentralization provides for participation of stakeholders, such as NGOs, local communities, tribal groups and complementary projects. It should promote local empowerment.
- The Kitanglad experience points out that decentralizing park area management does not merely mean devolving responsibilities previously concentrated within the national bureaucracy, such as the Department of Environment and Natural Resources. It involves the devolution of decision-making authority to various stakeholders.
- There is no ready template in protected area management. The Philippine experience demonstrates that sensitivity and recognition of cultural tradition and local knowledge, as well as flexibility to negotiate with various stakeholders, can sustain many of these local initiatives.
- Genuine and good community leadership produces effective community involvement and a good result. Therefore, the process of identifying leaders is a very critical undertaking in the area of decentralization.
- National government agencies need to harmonize laws with conflicting

provisions in order not to divide the community (see Chapter 7 on Indonesia's experience).

Note

The authors would like to thank the Center for International Forestry Research (CIFOR) and Elezia 'Bebit' Gozun (Department of Natural Resources) for making it possible for us to participate in the Interlaken Conference on Decentralization, Federal Systems of Forestry and National Forest Programmes in April 2004.

Chapter 17

State, Forest and Community: Decentralization of Forest Administration in Guatemala

Silvel Elías and Hannah Wittman

Introduction

The recent political decentralization of forest administration in Guatemala is a complex process involving a diverse set of actors and management practices. Communal and regional forestlands have been historically considered 'open access' by a state that has systematically ignored ancestral land and citizenship rights and destroyed local communities. However, local forest management has been effectively practised since the colonial period. Especially in Guatemala's Western Highlands, community groups have struggled to maintain access and control over their lands and natural resources, and defended communal property, traditional knowledges and customary rights, and local government systems. Government initiatives during the past decade have promoted the deconcentration of forest regulation to the municipal level. At the same time, local communities are demanding recognition and restitution of ancestral rights over land and natural resources, including national, communal and municipal forests. Community forestry in Guatemala faces many challenges, especially in the areas of social inequality and democratization. Experiences discussed here from several communities in the Western Highlands illustrate the local issues and institutional capacities in the current process of decentralization. For decentralization of the forest sector to fulfill its objectives of controlling deforestation and environmental degradation, it is necessary to establish a better balance of power, to empower local organizations and to increase capacity for local management regimes in communities and municipalities.

Decentralizing forest administration in Guatemala

In Guatemala, recent laws passed in favour of decentralizing forest administration seek to streamline natural resource use and conservation while increasing the participation of civil society in decision-making around forests. This process has emerged from an imposed structural adjustment programme combined with the signing of the 1996 peace accords, which put a formal end to 36 years of civil war. The peace accords included provisions that recognized the historical claims of local communities over land and natural resources. Despite recent advances in these areas, progress has been slowed by a lack of dialogue between local stakeholders and government agencies, and by the maintenance of powerful interests that seek to maintain control over forests and other natural resources. In practice, decentralization of forest management in Guatemala has taken the form of 'municipalization', or deconcentration, in which a highly centralized forest regulatory system has been delegated to municipalities.

The principal challenge for decentralization in Guatemala is not only to transfer new responsibilities and administrative functions to local levels, but also to fundamentally transform historical power relations that have long excluded local villages and communities from participation in forest policy. In the natural resource sector, this challenge proves to be doubly problematic because land, forests and other resources have historically been principal sites of social contention between local communities and government institutions, especially municipalities.

This chapter analyses the situation of local institutions for forest management in Guatemala's Western Highlands. Rural communities have preserved mechanisms that have served to protect the few remaining forest areas in the Western Highlands since ancestral times, despite adverse conditions of social exclusion, extreme poverty, high population density and extreme land concentration. In this chapter, we describe the historical diversity and complexity of forest management systems in the Western Highlands and evaluate the recent decentralization initiatives, asking the following questions. In what ways does the Guatemalan process of decentralization via municipalization of forest regulation take into account the effectiveness of local institutions? How can rural communities overcome historical situations of exclusion vis-à-vis municipalities to become important allies in the efforts to democratize forest management?

Struggle for control of land and forests

Guatemala's Western Highlands are one of the poorest areas in Latin America. Poverty rates exceed 93 per cent among indigenous populations, and population density is the highest in the country outside of Guatemala City. More than 95 per cent of farms in the region suffer from poor soils and are below subsistence size. Highland forests are important not only because they produce benefits for local populations, but also for their contribution to global systems, including protection of water sources and conservation of endemic species (Elías Gramajo, 1997).

Present-day deforestation and conflict over the Western Highlands' remaining forests stem from the agrarian policies imposed since the colonial period that have defined mechanisms of access, property rights and land use. The confrontation between indigenous populations who sought to defend their patrimonial rights and governmental insistence on creating municipal jurisdictions for social control of these populations (Tiu and García, 2002) generated a chaos expressed in agrarian conflicts. During the colonial period, the Spanish Crown appropriated all land, subsequently granting 'royal titles' to select indigenous communities. During the Liberal period at the end of the 19th century, the state pressured these communities to sell, rent or divide their lands. Despite the existence of the royal titles, land rights were lost or ceded to municipalities, which in turn passed use rights or titles to large coffee producers. Legal battles over these land rights continue to the present day. Despite provisions in the 1996 peace accords, which commit the state to recognition of traditional and communal land rights for indigenous communities, almost no such recognition has yet been achieved. Today, 70 per cent of the land is concentrated in just 7.9 per cent of rural properties (INE, 2004).

The link between poverty and environmental degradation has been well recognized in the literature (see Blaikie and Brookfield, 1987). This link holds true in privately owned areas in Guatemala, which are almost completely deforested. However, aggregate forest cover in the Western Highlands has been retained at a higher rate than other areas in Guatemala. Several studies have attributed this conservation to greater community participation in the preservation of communal forest areas administered by a social collective – that is, a village community or municipality (Veblen, 1978; Utting, 1993; Elías Gramajo, 1995, 1997; Katz, 1995; Secaira, 2000; Wittman and Geisler, 2005). Villages, in particular, continue to exert control over these areas based on internal norms, customs and regulations that are periodically renegotiated and reinforced to fortify customary law, physical boundaries and local social relationships. For example, local forest guards and village leaders (honorary rotating positions) police forest areas for infractions, oversee communal plantings, regulate harvests and maintain common lands for multiple community uses. Until the present day, management and administration have been differentiated between community and municipal entities depending upon local history, customary rights and circumstances. These mechanisms constitute the basis for locally sustainable management institutions, which have made possible the conservation of resources worldwide but often lack adequate recognition in the design of national and international policies (McKean and Ostrom, 1995; Ostrom, 2000).

Forest sector institutions

State initiatives to regulate land and forest use during the past 30 years have been based upon three principal laws and a variety of administrative agencies. At least four levels of decision-making for forest management exist, overlapping at different moments during any given decision process. The international level comprises international agencies and non-governmental organizations (NGOs) that pressure the Guatemalan government to adopt policies, agreements and global discourses (for example, structural adjustment, free trade, global climate change, biodiversity and sustainability). The influence of these international actors has been especially pronounced in the creation of parks and protected areas, which tend to emphasize the protection of environmental services (biodiversity and carbon sequestration) while failing to recognize the forest goods and services utilized by local populations.

A second level is represented by the governmental agencies that are linked to land and forests – the National Forest Institute (INAB), the National Council for Protected Areas (CONAP), the Ministry of the Environment and Natural Resources (MARN) and the Land Fund – and also to projects, networks, commissions, coordinating councils and technical assistance offices, whose policies change according to the current government's orientation. At a third level, 'intermediary institutions', including municipal governments and a variety of national or regional NGOs, act as extensions of the state or are used by the state to extend its influence or control towards civil society.

Finally, a fourth level includes grassroots organizations with varying foci and levels of integration. Most related to forest initiatives are the Indigenous Community, *parcialidades*, local associations, indigenous mayoralties, auxiliary mayoralties, local development committees, communal assemblies and other organizational forms. The Indigenous Community is a legal entity operating mainly in eastern Guatemala, adopted by local organizations to ensure autonomy over local territorial issues. *Parcialidades* are communal and kin-based organizations primarily oriented around community forest management in the Western Highlands. They and the local associations also have legal status; the others operate based on customary and ancestral rights. The lack of legal representation is often utilized by the state to de-legitimize these organizations, despite their effectiveness and contributions to sustainable forest management practices in the Western Highlands.

The four institutional levels are linked and result in a complex dynamic of scales, encounters, contradictions and cooperation related to forest management. In Guatemala, the decentralization process currently under way follows a top-down model, a process that tends to 'increase the power of the state at local levels, creating new organizational structures to control local populations' (Ferguson, 1994). In addition, institutional levels are cross-referenced by ever more powerful and diverse market pressures for forest goods and services.

Rights and responsibilities

What is the current direction of the decentralization process in Guatemala, given the complex framework of forest management institutions and the current emphasis on deconcentrating authority for forest management to municipalities? Is it enough to strengthen the technical and administrative capacity of municipal agencies? What is the role assigned to community organizations?

Faced with the failure and high costs of governmental forest management (PAF-G, 2000), the state is currently attempting a decentralization policy to organize and ensure a better use of forest goods and services. The recent rapid changes in, and implementation of, policy imply that no final evaluation can yet be made regarding the direction or impact of the evolution of the forest sector and community forest management institutions. However, considering the historical relations of power surrounding forest access and management, the complexity of the related institutional sectors and, above all, the concrete experiences of the communities who currently manage communal forests, several observations can be made to help understand and reorient the current process of decentralization.

The new decentralization model was initiated in 1996 in an attempt to dispel the image of bureaucracy and corruption inherited from previous government forest agencies. At the same time, the state sought new mechanisms to organize social relations around forests and to delegate specific responsibilities to municipalities, which were guaranteed a position on the director's council of the newly created National Forest Institute. A sub-agency, the Communal Forest Office, National Forest Institute (BOSCOM) was created to deal specifically with community groups, but since 1996 has developed relationships with just five communities. BOSCOM today prioritizes working with municipalities, especially in the creation of municipal forest units.

The National Forest Institute has focused on strengthening municipal capacity to the degree that municipalities function as extensions of the central government. Communal forests have historically been a 'headache' for the government institutions. As these forests represent practically the only remaining forest cover in the Western Highlands, communities, government and NGOs are all interested in their preservation. However, community organizations continue to have little confidence in the government's capacity to administer these areas since historically government entities have placed communal patrimony at risk. Scepticism and resistance to government initiatives remains high. For example, at the end of the 1970s, the US Agency for International Development (USAID) project *Bosques Comunales* (exercised through the now-defunct government forest agency INAFOR) pressured various communities of the Western Highlands to implement forest management activities based on timber extraction, arguing that such management would improve forest benefits and forest health. Communities saw the initiative as a means for external actors to appropriate local timber and other forest resources. As a result, the community of San Francisco el Alto in Totonicapán made the decision to prohibit the municipality from entering into agreements with any external entity (governmental or non-governmental) for communal forest administration, and resistance to government initiatives in the region has continued. Nonetheless, the communities continued with their own programmes of forest administration, including the development of tree nurseries and seedling production, reforestation, forest monitoring and harvest regulation, resulting in maintenance of forest cover to the present day.

CONFLICT, ADAPTATION AND RESISTANCE

The Western Highlands represents a rich diversity of community and traditional management experiences. Guatemala's 23 ethnic and linguistic indigenous groups account for more than 65 per cent of the country's population, and each community has developed a set of relations and experiences over time based upon local conditions, resources and history. Power relations and dynamics within communities are as diverse as the communities themselves. Women and the poorest sectors of society have historically had little formal participation in the traditional rotating authority system of communal governance. In practice, however, women have maintained their access to fuelwood and forest products, while men have been reserved the right to harvest timber. Women also have important roles in seedling production, reforestation and maintenance of communal areas for grazing. Issues of gender equity are at the forefront of discussion in several indigenous organizations in Guatemala and continue to be a subject of negotiation.

Guatemala's recent policies for deconcentration and decentralization of forest administration and management do little to recognize the diversity of local systems that are adapted to the history and local conditions of each community. Instead of integrating a process of strengthening local capacity and increasing participation for all sectors of society (such as women and the poor) in forest administration, the policies have created yet another layer of exclusion in which municipal governments and external organizations have assumed the right to regulate forest use and management.

Negotiations between local and municipal actors

One positive experience illustrates the potential of municipal–communal negotiations over forest access and administration where local rights and customs gain legal recognition. Acknowledging the importance of local actors in forest administration, the municipalities of San Juan Ostuncalco and Concepción Chiquirichapa in the Quetzaltenango region have used their capacities, newly awarded through the forest decentralization initiatives, to pursue collaboration agreements and reaffirm links to community groups. Here, local capacity is not only recognized by the municipal government; but efforts are also made to strengthen it. In San Juan Ostuncalco, a municipal office has been created to coordinate and support capacity-building activities with local groups, and in Concepción Chiquirichapa the municipality has signed an agreement with local farmers to regulate the harvest of forest litter, traditionally used as the principal organic fertilizer in potato cultivation.

However, such positive collaboration has not been the norm throughout the Western Highlands. The historical relationship between municipal and village-level authorities in the Quetzaltenango region has been characterized as more horizontal and balanced compared with other municipalities in the Western Highlands (Grandin, 2000). The main contradiction expressed in this region is that despite the growing recognition of local contributions and capacity for forest protection, state policies and actions have not always respected or strengthened these initiatives. Instead, local experiences abound in which the state has destroyed, annulled, failed to recognize or co-opted local institutions. Instead of recognizing and legitimizing local- and community-level capacity for forest management and protection, the Guatemalan decentralization process, in general, seeks administrative efficiency through the imposition of top-down models of forest control administered through municipal forest offices.

For example, governmental policy and National Forest Institute regulations now require communal forests to be submitted to licensed technical 'management', which clashes with local perspectives. Community members associate 'management' with timber harvest practices and clear-cutting, based on what they see occurring on private and government-managed forest areas. Communal forests, on the other hand, are conserved for fuelwood extraction (usually deadwood) and harvest of forest litter and other non-timber forest products. In addition, these forests serve as catchments for local water sources (perhaps the most valuable benefit of Western Highland forests) and as sites for sacred or religious activities. Government insistence on a specific kind of externally approved and licensed 'management' ignores local practices, knowledge, norms, rights, local governments and local arrangements or 'institutionalities' that have persisted for many generations and have sustained local forest management practices.

Adapting local administration models

Because of communication difficulties and lack of legal recognition, many local communities have resorted to creating or transforming themselves into legally constituted entities to better negotiate with municipal and state entities. Some of these 'new' organizations were established to represent the entire social collective, whereas others were created to represent special interest groups or small groups of power holders (Elías Gramajo and Reyna Contreras, 1999). For example, the *Parcialidad de Vicentes* was formed just 15 years ago (all the other *parcialidades* in Totonicapán date back more than half a century) as a local defence strategy to reaffirm territorial rights following a conflict with the state. Currently, this group continues to administer the forest in coordination with the traditional local governments, the *Alcaldía Auxiliar* and the *Asamblea Comunal*.

After many years of autonomous management, the *Parcialidad Baquiax* in the municipality of Totonicapán established links to BOSCOM in order to implement a forest management plan in the area. This *parcialidad*, like all others in Totonicapán, existed for the purpose of forest administration and had legal standing, statutes and even a registered land title for the communal forest holdings. To facilitate its work with the *parcialidad*, BOSCOM required the creation of a forest committee with higher standing and authority than the original leadership. This committee became an intermediary between the original director's council and assembly and the government agency. The internal forest committee emitted even more rigid regulations for access and forest extraction in order to comply with the BOSCOM forest management plan. These new regulations principally affected the poorest members of the *parcialidad* and members of the neighbouring village, who had traditionally been allowed to harvest certain forest products.

In similar cases in the area, grassroots organizations resisted compliance with rules imposed by outsiders. Realizing that they were putting their control over local processes at risk, the local organizations chose to continue operation based on their own collective initiatives (ADEGO et al, 2002).

IMPLEMENTATION OF NEW ENTITIES AND RULES

New models and regulations

As part of their alliances with state agencies and environmental and development organizations, several municipalities have adopted new forest administration models that centralize decisions at the municipal level, in addition to creating new administrative entities. For example, in the municipalities of Quetzaltenango, Totonicapán, Zunil and San Martín Sacatepequez, the *ProBosques* project has stimulated the creation of protected areas departments to facilitate implementing new rules and regulations in which local management is formally and completely suspended in favour of municipal administration.

Costs and benefits

In the municipalities of Sibinal, Tacaná, Ixchiguán and San José Ojetenam in the department of San Marcos, communal forests are administered by community groups that have continued to struggle for recognition of their ancestral rights to these forests. Informal agreements had long existed between municipal and village-level governments regarding use rights over municipal forest resources. Here, local control has historically resided not in the municipalities, but in community organizations that have been linked to the resources and have effectively protected forests by limiting access and use to local needs, including fuelwood, water protection, local construction and space for religious activities (Wittman, 2001; Reyna, 2002). With the recent creation of municipal forest offices as part of the decentralization process, decisions and licences for forest activities are now generated from these external offices. This measure has increased municipal power over the forest areas within municipal territorial boundaries and has also, in turn, severely weakened the capacity of local groups who historically have protected the forests, putting forest stability at risk in addition to limiting forest access for traditional users.

With the installation of municipal forest offices, municipalities have effectively increased administrative capacity and collected revenue from licensing fees authorized within their jurisdiction. However, these funds are rarely reinvested or shared with the communities who live in and near the forests. For example, in the municipality of Ixchiguán, each communal forest was linked to a specific local community group charged with control and administration. The establishment of a municipal forest office in Ixchiguán has fuelled the desire of municipal elites to take control over the remaining timber from the traditional village groups.

Such agreements are mutually beneficial in that the municipality obtains income and increases its jurisdiction, and the National Forest Institute is able to 'delegate' the implementation of politics and administrative instruments to the municipality. However, communities have been faced with increased responsibility (and accountability) for forest protection, while local benefits have been reduced, as well as local administration and negotiation power. For example, where permission for fuelwood extraction had traditionally been granted at the local or community level, municipal licences are now required even for subsistence use from communal forests, thus 'centralizing' control over these resources. At the same time, failure to obtain a licence for extraction thus criminalizes subsistence use of local resources. Local experience indicates that fewer than 10 per cent of local residents can obtain municipal licences, requiring travel in addition to literacy and money to pay the fees. Fearful of retribution for failing to obtain a licence, many families either harvest fuelwood at night or send women and children to collect the wood, a task that traditionally had been carried out by men. Usurping the authority of local forest guards in favour of municipal forest guards has reduced forest monitoring overall and led to an increase in illegal logging by outsiders.

New local associations

An example from recent history illustrates some possible outcomes – including corruption and co-option – of new intermediate layers. During the 1980s, an epidemic infestation of pine beetles (*Dendroctonus* spp) and restrictions imposed by the armed forces during the civil war led to an increase in illegal extraction of forest products by non-community members. Faced with rapid forest degradation, several external environmental organizations offered assistance but found that local groups lacked legal standing. The environmental agencies promoted the constitution of a new umbrella organization, *Uleu Ché Já*, in the municipality of Totonicapán, meant to serve as a parallel institution to the traditional representatives of the *Associaión de las Alcaldes Auxiliares* and the *Comité de los 48 Cantones*. The new organization quickly became a point of reference for community forest management initiatives, but soon lost credibility, suspected of being controlled by the municipality, which for years had openly opposed the recognition of local and community management rights.

Some local community organizations in the Western Highlands have recuperated from this state co-option, however, and traditional community groups have continued to function despite the political pressures and governmental intervention (Dary et al, 1998).

From communal forests to protected areas

Faced with the difficulty of implementing forest management plans and the growing interest in conservation, the future of the Western Highlands communal forests may be increasingly directed towards newly created protected areas, moving even further away from a decentralized or local communal forest management. Promotion of such areas dates back to the 1960s, when the National Park Maria Tecún was created to protect the communal forests of Totonicapán. The highlands' many volcanoes were next declared parks; more recently, protected areas were designed around Lake Atitlán, the Sierra de los Cuchumatanes and the Visis Cabá Biosphere. However, in almost all of these cases government implementation never left the planning stages.

Since 1997 several municipalities – Tecpán Guatemala, Quetzaltenango, Zunil, Totonicapán y San Martín Sacatepequez, Cuilco and Tacaná – have been converting communal and municipal forests within their jurisdiction into protected areas, with the support of environmental NGOs. International organizations have shown increasing interest in Western Highland forests, including the World Bank with the Integrated Management of Natural Resources in the Western Highlands (MIRNA) project; the Inter-American Development Bank with the Prioritization of Strategic Areas in the Western Highlands (PARPA) project; and projects under the Global Environmental Facility (GEF)/United Nations Development Programme (UNDP).

However, government forest protection measures have had harmful effects in certain cases. For example, during the 1980s, the state sought to protect the pinabete tree (*Abies guatemalensis*), one of the Western Highlands' endemic species that is severely threatened by commercialization as a Christmas tree. The measure was a typical example of responsibility transfer and reduction of local benefits: the state totally restricted any local use of this traditional resource, but required communities to participate in its protection, even though many community organizations already controlled use of this species. With the new legal measure, many communities abandoned their traditional protection systems since they no longer were allowed to use the resource, and the consequence was an increase in illegal harvesting. This example raises the issue of the future role of rural communities in the local administration of their traditional natural resources, and the benefits that local communities will gain through the new initiatives.

Conclusions and lessons learned

The new powers assigned to municipalities by the 1996 Forest Law, the creation of municipal forest units and access to forest incentives by municipalities and communities have been cited in a recent Center for International Forestry Research (CIFOR) study as positive examples of the decentralization of the forest sector in Guatemala (Ferroukhi and Echeverría, 2003). However, little attention was given to the impact of these measures on existing local institutions and administration programmes.

Local institutions in Guatemala have played an important role in the management and administration of local and communal forest resources. Strengthening municipal forest management programmes is important for decentralization in Guatemala but does not guarantee authentic decentralization to traditional user communities. Currently, the decentralization programme does little to support the rights of local institutions linked to communal forests, and it actually weakens systems of communal management that have protected the few remaining blocks of Western Highland forest.

Rural communities have little chance to participate in state-subsidized reforestation and forest management programmes. Strict government requirements for participation (including legal land titles and legal standing) are out of reach for many communities. The subject of land title, for example, continues to be taboo in many parts of the Western Highlands, where communities historically have struggled with the state to protect their land rights (Lovell, 1985; Davis, 1997; Palma and Taracena, 2002; Grunberg, 2003). The required elaboration of a costly technical forest management project is another barrier. Finally, projects that must conform to management practices approved by the National Forest Institute generally do not make allowance for local needs.

Financial incentives could potentially be useful to local communities not only to finance forest programmes, but also to stimulate other productive activities and strengthen local organizational capacities and social equity programmes. For example, the Association of Eulalensis Women for Pixan-Komop Development (AMEDIPK), a women's group in Huehuetenango, has achieved recognition for its forestry activities; another association in Sololá has successfully based its employment and income strategy on forest management. However, additional study is needed to better understand how these forest resources contribute to improving living conditions, social equity and local administration. Caution must be taken to ensure that resources are not used to consolidate the power of existing traditional economic and political elites.

The focus on strengthening municipal forest administration capacity has increased the power of the municipalities to generate income (via licensing and sales of municipal forest products) and to exercise control over all forest resources within their territorial jurisdiction. It has also extended state control over these natural resources, as municipalities have historically represented centralized power. The process of decentralization has been severely limited, however, by the failure to develop instruments to implement decentralization as a national strategy. In the forest sector, current programmes resulting from the 1996 Forest Law assign responsibilities to municipalities but are not linked to other decentralization initiatives, such as those emanating from the more general decentralization laws passed in 2002 requiring the implementation of urban and rural development councils.

Increased capacity for forest management at the municipal level does not necessarily mean an increase in the capacity at the local level. The subordination of community organizations to municipal agencies is counter-productive within a framework of policies designed to improve the quality and benefits of forests for local development. Local institutions in the Western Highlands have made important contributions to the conservation and management of remaining forests in the region; but this contribution has not been sufficiently recognized in the decentralization policies. How can local capacity for forest management be recognized and strengthened through a decentralization programme? We recommend the following:

- Regularize and legalize communal land tenure based on ancestral rights to guarantee community control over forest use and administration.
- Develop community land-use planning programmes in coordination with state agencies. This may be a challenge. In addition to issues of power inherent in such negotiations, the state views land-use planning as a technical exercise rather than as a social and political process.
- Create a dialogue about how traditional forest management practices and knowledge can be integrated with technical forest management practices promoted by state and non-governmental agencies. At the global policy level, increasing recognition and legitimization of local knowledge and management systems should facilitate discussion on how to integrate these objectives.

Within the framework of the proposed decentralization of forest and natural resources management, governmental and related agencies should establish mechanisms to recognize and support local administration carried out by local institutions, adapting these mechanisms to local specificities and strengthening local organizations. Such measures would serve to consolidate social capital and implement the proposals of grassroots organizations, establishing clear rules of interaction but not subordinating local governance structures to municipal authorities and other governmental, political or economic agencies. Only in this way can a truly effective and democratic decentralization occur for Guatemala's forest administration system.

Note

Research support and funding for this work came from Facultad Latinamerica de Ciencias Sociales (FLACSO) Guatemala; the Facultad de Agronomía at the Universidad de San Carlos (USAC), Guatemala; the National Forest Institute (INAB); CARE-Guatemala; the Cornell Institute for International Food and Development (CIIFAD), US; and the Mario A. Einaudi Center for International Studies at Cornell University. We are also grateful for comments from Violeta Reyna and Charles Geisler, and from anonymous reviewers.

References

ADEGO/AMEDIPK/CASODI/MAGA/RUTA (2002) *Poder Local, Participación y Descentralización: Lecciones aprendidas desde la perspectiva y experiencia de organizaciones locales*. Unidad Regional de Asistencia Técnica, Serie de Publicaciones RUTA: Sistematización de Experiencias No 16; also available from www.ruta.org/espanol/publicac/sistemat/pod-local-16.pdf

Blaikie, P. and Brookfield, H. (1987) *Land Degradation and Society*. Methuen, New York

Dary, C., Elías, S. and Reyna, V. (1998) *Estrategias de Sobrevivencia en Ecosistemas Frágiles. Los Chortís en el Oriente de Guatemala*. FLACSO, Guatemala City

Davis, S. (1997) *La Tierra de Nuestros Antepasados, Estudio de la Herencia y la Tenencia de la Tierra en el Altiplano de Guatemala*. CIRMA, Guatemala City

Elías Gramajo, S. (1995) 'Tenencia y Manejo de los Recursos Naturales en las Tierras Comunales del Altiplano Guatemalteco'. Paper presented to the XIX International Congress/Latin American Studies Association (LASA) 95, Washington, DC

Elías Gramajo, S. (1997) *Autogestión Comunitaria de Recursos Naturales: Estudio de Caso en Totonicapán*. FLASCO, Guatemala City

Elías Gramajo, S. and Reyna Contreras, V. (1999) *Comunes y Cacicazgos: El control de los Bosques y Territorios en Totonicapán*. FLACSO, Guatemala City

Ferguson, J. (1994) *The Anti-Politics Machine: 'Development', Depoliticization and Bureaucratic Power in Lesotho*. University of Minnesota Press, Minneapolis

Ferroukhi, L. and Echeverría, R. (2003) 'Las Políticas de Gestión Forestal Descentralizada en Guatemala', in Ferroukhi, L. (ed) *La Gestión Forestal Municipal en América Latina*. CIFOR–IDRC, Bogor, Indonesia

Grandin, G. (2000) *The Blood of Guatemala: A History of Race and Nation*. Duke University Press, Durham, North Carolina

Grunberg, G. (2003) *Tierras y Territorios Indígenas en Guatemala*. FLACSO/MINUGUA/CONTIERRA, Guatemala City

INE (Instituto Nacional de Estadísticas) (2004) *Informe del Censo Nacional Agropecuario*. INE, Guatemala City, www.segeplan.gob.gt/ine/index.htm

Katz, E. G. (1995) *Guatemala: Land Tenure and Natural Resources Management*. The World Bank, Latin American and the Caribbean Regional Office, Washington, DC

Lovell, W. G. (1985) *Conquest and Survival in Colonial Guatemala: A Historical Geography of the Cuchumatán Highlands 1500–1821*. McGill-Queen's University Press, Kingston, Canada

McKean, M. and Ostrom, E. (1995) 'Bosques Comunales en Propiedad Común: ¿Solo un Vestigio del Pasado?'. *Unasylva*, vol 46, pp3–15

Ostrom, E. (2000) *El Gobierno de los Bienes Comunes, La Evolución de las Instituciones de Acción Colectiva*. Fondo de Cultura Económica, México City

PAF-G (2000) *Guía para el Establecimiento de Oficinas Municipales de Administración de Recursos Naturales/Forestales en Guatemala*. Plan de Acción Forestal para Guatemala, Guatemala City

Palma, G. and Taracena, A. (2002) *Procesos Agrarios desde el Siglo XVI a los Acuerdos de Paz*. MINUGUA, Guatemala City

Reyna, V. (2002) *Autogestión en los Bosques del Altiplano Marquense*. FLACSO, Guatemala City

Secaira, E. (2000) *La Conservación de la Naturaleza, el Pueblo y Movimiento Maya, y la Espiritualidad en Guatemala: Implicaciones para Conservacionistas*. PROARCA/CAPAS/USAID, UICN, FCG and the Nature Conservancy, Guatemala City

Tiu, R. and García, P. (2002) *Los Bosques comunales de Totonicapán, Historia, Situación Jurídica y Derechos Indígenas*. FLACSO–MINUGUA, Guatemala City

Utting, P. (1993) *Trees, People, Power: Social Dimensions of Deforestation and Forest Protection in Central America*. UNRISD/Earthscan, London

Veblen, T. T. (1978) 'Forest preservation in the Western Highlands of Guatemala'. *The Geographical Review*, vol 68, pp417–34

Wittman, H. (2001) *Fragmentación y Manejo de Terrenos Comunales y Municipales en el Altiplano Occidental de Guatemala: Experiencias de las Municipalidades de San José Ojetenam y Ixchiguán, San Marcos*. CARE, Guatemala City

Wittman, H. and Geisler, C. (2005) 'Negotiating locality: Decentralization and communal forest management in the Guatemalan Highlands'. *Human Organization*, vol 64, no 1, pp62–74

Chapter 18

Decentralization: Issues, Lessons and Reflections

Doris Capistrano and Carol J. Pierce Colfer

Introduction

Decentralization of forest governance has been defined and implemented in different ways in a variety of contexts around the world. It has had varying results and impacts both on forests and on different groups of stakeholders with interests in these resources. Although the details may vary, countries' experiences with decentralization include some remarkably similar patterns, particularly in the aspects that appear to contribute to the success or failure of decentralization efforts.

The international workshop on Decentralization, Federal Systems of Forestry and National Forest Programmes at Interlaken, Switzerland, in April 2004 featured and shared lessons from different countries' experience with decentralization of forest governance. This conclusion distills some of the main issues and insights woven through the book's chapters and captures important ideas highlighted during the various sessions of the workshop. It summarizes the patterns that have emerged and reviews how decentralization looks in practice, as opposed to in theory. The chapter also briefly reviews the implications of decentralization for forests and biodiversity conservation. Finally, the chapter notes common challenges and concludes with lessons and observations.

Decentralization in practice

Perhaps one of the most dramatic findings emerging from these analyses and observations is the gap between the theory and the practice of decentralization. This is clear both from the multi-country analyses (see Chapters 2–4) and from the individual cases (for example, Chapters 7, 11, 13 at the national scale, and Chapters 15 and 17 at the community level). Whereas the ideals of decentralization – increased voice for local communities, greater accountability in local governments, more appropriate policies – are heard over and over again, the reality is quite different. The same unhelpful patterns emerge again and again: inadequate resources to accomplish the goals of decentralization; unwillingness on the part of those in power to cede significant authority and resources; uneven local capacities and will; elite capture; and unclear guidelines and division of labour among governmental levels. Democratic decentralization – that form characterized by 'the transfer of authority to representative and downwardly accountable actors, such as elected local governments' (see Chapter 2) – requires that representative and accountable local actors have autonomous, discretionary decision-making spheres with the power and resources to make significant decisions pertaining to local people's lives (Ribot, 2002). Certainly, in newly decentralizing countries, such a pattern is not in evidence; democratic decentralization remains a pipedream.

Both Sasu (Chapter 11) and Nsita (Chapter 10) on Ghana and Uganda, respectively, provide graphic descriptions of the problems their countries have in transferring resources from central levels of government to lower levels. Both countries have seen their decentralization efforts hamstrung by the lack of funds and other resources at lower levels despite policies and legal frameworks that would seem conducive to such transfers. Even more extreme are the problems transferring funds, authority and rights to local people.

But decentralization efforts are processes, and processes take time. Both citizens and government officials need time to work out the quirks in new systems, and many of the efforts described in this book are still in the formative stages. Indonesia, for instance, began its serious decentralization efforts only in 2000 (legalizing decentralization in January 2001), and it started with an unusually centralized system. Not surprisingly, then, there is a great deal of uncertainty about how to proceed at all levels of government and in communities as well. Such uncertainty is a fertile breeding ground for conflicts and adverse impacts of all kinds – among the citizenry, between citizens and government officials, and among governmental levels.

We may tend to think of adapting to decentralization as a linear process. However, the cases from Uganda and Ghana, as well as Australia and the US, demonstrate clearly that the process can be cyclical. All of these countries have gone through decentralizing and recentralizing processes over time. Although such evidence can be discouraging for those who see decentralization as a 'silver bullet' for the world's problems, in fact, it simply reflects another important lesson: there is no stability in governance. Because conditions, norms and the citizenry change, governance – and, thus, also decentralization – is an ongoing process of negotiation and power brokering.

Several participants noted that the forest sector often lags behind other sectors that are decentralized, such as health and education. In some cases, this seems to reflect the comparative unimportance of the forest sector; in other cases, such as in Ghana and Indonesia, quite the opposite applies: the forest sector is seen as too valuable for powerful people to relinquish. There is a marked tendency for central governments to decentralize management responsibility for the most degraded and least valuable forests while keeping the most valuable revenue-generating forests under central control. Thus, it is not just a coincidence that most community forestry, co-management and other forms of participatory forestry first gained a foothold on degraded forests and areas considered unproductive wasteland by the forest bureaucracy.

Scotland (see Chapter 12) and Zimbabwe (see Chapter 15) dramatically illustrate the importance of both top-down and bottom-up forces in making decentralization work. Decentralization seems to proceed most smoothly when an action at one level meets a supportive, responsive action at the other level. When different levels work together, much can be accomplished – and accomplished more easily. Center for International Forestry Research (CIFOR) research confirms this observation: in our work with communities using an adaptive collaborative management approach, we began at the local level but concluded that we need to work at higher levels of the government as well.

The lack of technical, institutional and other types of capacities has been consistently cited as a weakness and bottleneck in countries' efforts to decentralize. Governments and forest management bureaucracies have often used capacity deficiency at lower levels in the hierarchy as an argument against implementing decentralization and devolution. Likewise, local governments have resisted pressures for further decentralization to communities or village-level institutions, citing their lack of capacity and inability to manage forest resources effectively. The question is whether capacity-building should come first, or direct experience at handling decentralization. In Chapter 4, Ribot argues persuasively that officials at intermediate levels of government cannot gain the capabilities they need until they have the opportunity to deal with the problems that decentralization poses. Others, such as Ferguson and Chandrasekharan (Chapter 3), and Siswanto and Wardojo (Chapter 7), argue that the problems with decentralization exist precisely because people at the lower levels of government were not prepared to take on the responsibilities they were given. It is really a chicken-or-egg question, and we are unable to determine a 'right' answer. As with so many other issues in this field, the appropriate answer for any individual case depends upon the context – as many have said, in this domain, one shoe *definitely* does not fit all. One issue, noted by Larson (Chapter 2), is whether central government personnel remain available to those at lower levels when responsibilities are transferred. In some countries, central-level personnel are withdrawn, leaving lower-level officials unprepared and unsupported; in other cases, local-level officials themselves reject offers of help, thereby reducing their own effectiveness.

What are the main pitfalls and stumbling blocks for decentralization? Elite capture – that is, the ability of those with power and wealth to take advantage of new opportunities and enhance their existing power and wealth – is a recurrent problem. In many countries, corruption plagues efforts to improve governance and resource management. Weak civil society – difficulties acting collectively towards common goals – allows the powerful to continue acting in ways that do not serve the general interest. Such problems are exacerbated in societies that are separated by strong tribal or ethnic divisions, where institutional links among groups are rare. This in turn makes it difficult for citizens to band together to hold those in power accountable. Strengthening civil society seems to be one of the more probable entry points for making decentralization work as its proponents envision.

The problems at the village level in efforts to decentralize have already been mentioned (see, in particular, the chapters in Part III on community voices); but such problems are even more extreme when we consider the female half of the population. Women in most cultures have not been actively involved in political life, except at the very local (often sub-village) level. The same can be said of other marginalized groups, such as lower castes in the Indian subcontinent, hunter-gatherer groups in humid tropical forests, settlers in the Amazon, children and youth. None of the chapters in this book address such equity issues directly; but they should be borne in mind as we work towards a more equitable world.

Many practitioners and scholars concerned with decentralization see democratization as a major outcome to be expected from the process of decentralization. Given the reality of decentralization in practice, a major challenge is how to foster processes that are inclusive and sufficiently flexible to adapt to different situations and, at the same time, enhance democratic and accountable governance. But this raises another important question. Many of the analyses assume that the Western model of democracy is suitable and desirable for the whole world. The emphases on accountability, transparency and 'one man, one vote' are straight out of the West. Yet, some reviewers of these chapters have noted that different cultures approach these political issues from different perspectives. Some have argued that accountability, transparency and voting are alien imports and may be inconsistent with local cultural norms (for example, Wollenberg et al, 2004). This argument has also been made with regard to women and their involvement in formal political processes. Sithole (2004) has argued that rural Zimbabwean women prefer their informal, behind-the-scenes approaches to political influence (rather than the usual mute 'participation' in formal committees and local government structures). Like so many issues related to values, there is no easy answer; but this is an issue that may bedevil some decentralization efforts.

FEDERAL SYSTEMS: PATTERNS AND INSIGHTS

Since forests and natural resources are typically managed in a decentralized manner in federal systems, the Interlaken workshop focused on federal countries as a possible source of experience and insights for contemporary decentralization efforts. However, any insights have to be interpreted with a lot of caution. There is, in fact, a fundamental difference between decentralization in federal versus unitary systems of government. Federal systems actually represent a case of reverse devolution – rather than central authority being devolved to a lower level, separate entities come together in some sort of federation, ceding some of their power to the central authority.

In federal systems, federation members, as historically decentralized political and resource management units, have agreed through their national constitution to confer certain responsibilities and authorities to the central government in pursuit of their common interests. Thus, instead of decentralization and devolution, Gregersen et al (see Chapter 1) argue that the operative principle in federal systems is constitutional non-centralization. In federal systems, the meso and local levels of government are often well established, with longstanding political constituencies and accountability mechanisms. They tend to have real, constitutionally defined rights and authorities in contrast to meso and local levels of government in unitary, non-federal systems. These latter are not constitutionally empowered, but rather are subordinate units of the central government.

Based on their review of eight federal systems (Australia, Canada, Brazil, India, Malaysia, Russia, Nigeria and the US), Gregersen et al observe that different federal systems have had varying degrees of success in realizing the benefits and minimizing the costs of decentralization. This is hardly surprising given the wide contextual variation among federal systems, in general, and in the specific countries examined. The countries studied vary in many respects: in forest-related authority and responsibilities; in the fiscal responsibilities assigned to different levels of government; in the patterns of forest ownership (with private ownership predominating in the US, for example, and state ownership predominating in India); in their policies and institutional structures to deal with civil society and the private sector; and in their cultural predispositions regarding authority, freedom, equity and other core values that define political norms.

Nevertheless, one common experience among these federal systems is the apparent difficulty each has had in decentralizing to the lowest level of government. This is a familiar pattern observed in decentralization in unitary systems as well. Despite their better-articulated and better-developed meso levels of government, federal systems evidently still tend to be inadequately prepared or involved in mediating between the local and central levels. Even where there is a clear policy to decentralize to the lowest level, as in India with its joint forest management programme, lack of capacity of local institutions and undeveloped mechanisms to ensure competent management and accountability at different levels have tended to impede decentralization to the local level. However, in some cases, such as in Bolivia (see Chapter 9), despite limitations, some municipalities have been able to serve as effective interlocutors and conduits between communities and central governments. The renegotiation among entrenched vested interests at different levels, including the bureaucracy, is an important component of this dynamic and determines how much power and resources are actually shared with communities and other actors in a decentralized system. This renegotiation tends to be particularly difficult, and the downward flow of power and authority is less likely when high-value resources are involved.

Although decentralizing to the local level is clearly a challenge common to both federal and unitary systems, Gregersen et al (Chapter 1) observe that it tends to be more difficult in countries where local government capacities, revenue management and accountability mechanisms are less developed. Moreover, local governments may not have enough incentive to take on their assigned forest management responsibilities since they are often not provided commensurate rights, authority and resources with which to perform their new roles. Many districts in the Philippines, for example, have attempted to defer the handover of forest management from the central government to the districts, citing lack of clarity in defining responsibilities and inadequate funding, facilities and staff, as well as complexities in administrative arrangements (see Chapter 3).

Tussles over jurisdiction and authority between the national and provincial or state governments are not uncommon even in federal systems. Just as in unitary systems of government, the roles, responsibilities and balance of power and authority among layers of government in federal systems are not static. They are continuously tested and renegotiated and need to be adjusted over time. Perhaps partly as a result of this ongoing adjustment, Gregersen et al noted a general lack of clarity and wide discrepancies between the official and actual distribution of power in many countries.

In examining the decentralization experiences of different countries, including both federal and unitary systems, Ribot (Chapter 4), Larson (Chapter 2), Ferguson and Chandrasekharan (Chapter 3), Gregersen et al (Chapter 1), Pacheco (Chapter 9) and other authors see the critical importance of the dynamic balance between authority, accountability mechanisms, responsibilities and revenue-sharing across different levels of government. Experience from diverse countries indicates that the relationship and balance among these elements determine, to a large extent, the effectiveness and efficiency of decentralized systems of forest governance. A successful framework for decentralized forest governance requires at least three things:

1 appropriate and effective sharing of authority to make decisions and raise revenues, and sharing of responsibilities among levels of government according to their individual abilities and needs;
2 effective enforcement and accountability at all levels of government to ensure that government agencies are acting fairly, efficiently and effectively in carrying out their mandates; and
3 effective linkages with other sectors that affect or are affected by the forest sector.

Key aspects of each of these ideas are reflected to varying degrees and are analysed in different chapters in this volume.

Not unexpectedly, countries and institutions go through a learning curve. Longstanding federal systems tend to have developed institutionalized mechanisms to deal with the continuing disagreements among their central and sub-national units, in contrast to more recent federal systems that still struggle with the complexities of the process (see Chapter 1). This suggests that countries now undergoing decentralization can learn just as these longstanding federal systems have, and their institutions can be expected to adjust and function more smoothly over time as they gain greater experience and confidence with decentralization.

Australia provides a good example of such institutional learning and adjustment in recent years. Ferguson and Chrandrasekharan (Chapter 3) recount major disputes about the granting of export licences and environmental considerations, which pitted national stakeholders – principally the forest industry, unions, landholders and non-governmental organizations (NGOs) – against provincial stakeholders during the 1980s. In this case, the tussle appears to have spawned an institutional innovation during the 1990s, leading to the creation of a joint national–provincial agreement, the Regional Forest Agreement process. This is an institutional mechanism to mitigate potentially destructive differences between national and provincial governments. Similar types of forums and mechanisms exist in other countries. In Malaysia, the National Forestry Council serves as a forum for federal and state governments. In Canada, the Canadian Council of Forest Ministers is focused on making more effective and efficient linkages between federal and sub-national governments and communities. A range of other mechanisms, including national forest policy statements and national forest programmes and processes, are also evolving in different countries to address conflicts or facilitate coordination among levels of government. Together, these can be the backbones for more effective decentralization in the future.

The learning curve of countries that are only beginning to embark on decentralization can be made easier and more efficient with the benefit of lessons and insights from countries that have a longer experience with the process. It is exactly for the purpose of facilitating the sharing of experiences and lessons among countries that the international workshop on Decentralization, Federal Systems of Forestry and National Forest Programmes was conceived and organized. As a major forestry institution, the United Nations Forum on Forests (UNFF) was also envisioned as a place where the distilled lessons and insights from these exchanges could be more widely publicized, disseminated and, perhaps, also adopted for practical follow-up action.

Decentralization impacts on forests and biodiversity

One of the common assumptions made about decentralization is that encouraging local participation and more equitable sharing of benefits from forest management at the local level will foster more sustainable use and management of forest resources. Although there are many cases of forests' being better protected or rehabilitated after handover to local control and management (for example, the Philippines, India, Nepal and Guatemala in Chapter 6), decentralization of forest management can also lead to ecologically unsustainable outcomes. In certain cases, where there are immediate trade-offs between resource conservation and local development, decentralization and devolution of rights and management authority to local institutions may even facilitate local choices in favour of short-term development options to the detriment of forests. In Indonesia, for instance, local institutions tend to be weak, corruption is rampant, economic returns to forest resource exploitation are high, and there is considerable uncertainty about decentralization and its longevity as a policy. In some districts, decentralization has led to increased logging and forest loss in the short term. Stakeholders' uncertainty about their continuing rights to forest resources is a major driver of such unsustainable use.

Russia, following the collapse of the Soviet Union, provides another example of adverse environmental impacts from incomplete and hasty decentralization. After 200 years of mostly highly centralized state control of the forest estate, in which local forest management units were merely executors of tasks without decision-making authority, Russia transferred logging responsibilities for 98 per cent of its former timber industrial complex to the private sector, and decentralized forest management responsibilities to regions and local management units. What followed is a familiar story: regional and local units, inadequately funded and ill prepared for their new roles, increased selective logging to raise needed local revenues. They also under-invested in fire protection, replanting and overall management. Incidence of disastrous fires increased, and forest management deteriorated. Having lost management authority and control over forest revenues, the central government, which remained responsible for forest pest control and research, also lost the capacity and the incentive to continue to perform these roles. Citing similar patterns of forest destruction from two previous episodes of nationalization and local management of forests, after the 1917 October Revolution and under Krushchev's reforms in 1957–1966, Malysheva (Chapter 13) is not optimistic about the current decentralization. She expects that forest resources will be exhaustively exploited for foreign exchange by regional administrations and private owners and argues that centralized forest administration works much better for Russia's forest estate. Other economies in transition share the same concerns; however, they generally see decentralization as an important tool to facilitate a shift from centrally planned management to a market economy. Faced with similar dilemmas, and torn by divergent impulses to decentralize or return to centralized control, they would rather tackle forest reforms through political consensus, cooperation and clear and equitable division of responsibilities between national and sub-national authorities. They also note the importance of having financial resources and clear systems of accountability, legal frameworks that reflect emerging trends and needed reforms, and authority for forest use, protection and renewal.

Indeed, both in the chapters and in discussions during the workshop, participants raised concerns about the likelihood of forest deterioration following decentralization. Yet, there are also examples in which, after the initial round of resource destruction, local stakeholders and institutions have organized themselves to better manage degraded or rapidly dwindling forest resources, with numerous examples from India (Orissa, Gujarat, West Bengal and Kumaon) and Nepal. With the necessary social capital and institutional capacity, leadership, motivation and incentives, the initial phase of forest destruction can be, and has been, reversed through concerted local and societal action to protect remaining forests and to rehabilitate those that have been degraded.

An example is Switzerland's cycle of decentralization, forest degradation and rehabilitation during its 150 years of decentralized forest management. Deforestation increased during the first 20 years following decentralization, prompting recentralization of some management authority, along with the development of needed technical and institutional capacities. The current decentralized system of Swiss forest governance has been shaped for more than a century by the process of defining an appropriate balance of shared roles, responsibilities and technical expertise between central government and cantons. In terms of biodiversity conservation, in the case of Switzerland, decentralization and community forest management have proven in the long term to be far superior to private forest management.

One of the major concerns about decentralization is its tendency to fragment management responsibility over landscapes that are better managed together. This issue is particularly a concern for protected areas and areas of high biodiversity value. In most countries, precisely because of this concern, protected areas and forest reserves have largely been exempt from decentralization and have remained under central government control. There was agreement among participants that although forests have important values for local communities who are favoured under decentralized management, it is also important to ensure the maintenance of other public goods – notably biodiversity and hydrological and climate regulation functions – which benefit the national and global community. This may imply improvements in law enforcement at various levels to reduce negative impacts deriving from self-interest and resulting unsustainability.

Many protected areas and conservation reserves have tended to be focal areas of conflict between local people and forest management authorities. Based on the dominant paradigm at the time, many present-day protected areas and conservation reserves came about through gazettement or unilateral acts of colonial or post-colonial states. Many were created through the removal of important natural areas from local use and control, in effect appropriating and placing them under the 'highest competent authority' of the state (see Chapter 6). This process served to disenfranchise many local, traditional and indigenous people and their systems of knowledge about forest management, and instead vested management authority in the state, whose system of technical knowledge and practice was affirmed as superior and legitimate. Conflicts between local people and forest agencies in and around protected areas are often rooted in the history of their establishment. In many cases, these are conflicts not only about forest resources, but also about competing systems of forest knowledge and values. In many areas, people's traditional uses of areas long considered their own are defined by governments and protected area managers as encroachment, poaching and illegal logging; local people can become de facto criminals overnight (see, for example, Chapter 17). In addition to a history of animosity and mistrust, such conflicting views, in which the definitions of governments and protected area managers typically carry far more weight than do those of local people, can result in local people's refusal to cooperate in the areas' protection.

Mainstream environmental groups still advocate keeping at least 10 per cent of all forests as inviolate protected areas. However, during the past few decades, there has also been increasing support for the idea of involving local people in managing conservation programmes, particularly where win–win solutions to competing agendas of conservation and development can be found. While conceding that well-conceived and well-executed decentralization has resulted in more extensive and mixed forests and has yielded real biodiversity benefits, Sayer et al in Chapter 6 are measured in their views on decentralization, noting that there is no single correct formula, and that the degree to which biodiversity conservation can be devolved or decentralized depends very much upon societal choice.

Their review of selected cases provides both positive (Bolivia) and negative (Indonesia), as well as mixed (Peru), results. However, from their assessment of integrated conservation and development projects, Sayer et al find very few cases of real devolution of rights and assets to local people and few examples where the benefits exceed the opportunity costs of the protected area to local people. This assessment is consistent with the overall conclusion of other authors in this volume, notably Larson (Chapter 2), Ribot (Chapter 4), Ferguson and Chandrasekharan (Chapter 3), and Elías and Wittman (Chapter 17), about people's participation in decentralized forest management initiatives. Cases of significant employment and livelihood creation from people's involvement in nature parks also tend to be rare. Sayer et al (Chapter 6) provide several promising examples of local people's involvement in livelihoods and conservation-oriented projects in Guatemala, Nicaragua, the Philippines and Eastern Africa. While there appear to be initial indications of positive biodiversity conservation results, as they point out, it is unclear if these can be sustained. Assessments of biodiversity across sites are notoriously difficult; this complicates making comparative assessments of decentralization impacts across sites and governance approaches. Funding is another factor constraining the ability of local people to formulate management plans, access needed skills and sustain collaborative interactions with other stakeholders – all necessary for sustained, effective forest management and conservation (see Chapters 15 and 16).

How central and local levels of government share management authority and responsibility, as well as benefits from protected areas and other conservation forests, must be determined on the bases of both local interests and broader public goods considerations. Equity in the sharing of benefits and burdens, and respect for alternative systems of value and knowledge will be important in reducing tensions and conflicts over these areas. Building on local knowledge and strengthening local institutions have proven to be invaluable and essential for effective conservation results and meaningful local participation (see Chapter 16). Monitoring and assessment processes using sound approaches and reliable technical and management information could facilitate broader learning and enhance the credibility of results and insights from decentralized conservation initiatives with local people's participation.

During the workshop, participants agreed that centralized approaches to protected areas should target sites of national importance, and that central authorities also need to determine the extent to which rights and access to high-value areas can be decentralized (Report to UNFF4, 2004). At the same time, participants recognized the limitations of decentralization and noted that measures are required to ensure that decentralization does not, in fact, lead to fragmented and dysfunctional landscapes. There was agreement that central institutions often need to play a role in providing a spatial context for local actions and in setting the limits within which decentralized managers operate.

Participants also agreed that to take advantage of the potential benefits from decentralization and to guard against potential negative impacts, countries would be wise to observe the following principles (Report to UNFF4, 2004, p14):

- Whenever possible, traditional communal forest management systems, which provide multiple functions, should provide the basis for decentralization. Such systems are likely to be effective in meeting local needs, better adapted to local conditions and resilient to external influences.
- Markets need to be developed for environmental services (particularly water protection, climate change and biodiversity), based on secure property rights, in order to provide the revenue support for the provision of those services and as a more equitable way for society to exert influence over which national and global values are delivered.
- Centralized approaches to protected areas should target sites of national importance, and any local opportunity costs of such areas should be compensated in an adequate way. Additional ecological values should be conserved at a landscape level through decentralized multifunctional management systems.
- Central institutions should use participatory approaches in establishing the limits within which decentralized systems operate. They need to provide the spatial planning context, define the permanent forest estate and otherwise support regulatory and incentive frameworks.
- Economies of scale tend to favour uniform approaches in large-scale centralized schemes for the restoration of degraded lands. With the right

framework of incentives and property rights, decentralized systems can favour more biologically diverse and locally adapted approaches to restoration.

COMMUNITY PERSPECTIVES ON DECENTRALIZATION

The context for community involvement and participation in decentralized systems of forest governance has been slowly undergoing profound structural changes over the past two to three decades. As a result of previous – mostly experimental and mostly external – donor-funded programmes on social forestry, participatory forestry, integrated community-based resource management and similar initiatives, the concept of local people's and community participation in resource management is now a part of mainstream consciousness. Unfortunately, policy support for this concept in many countries remains at the rhetorical level, even in countries where decentralization is official forest policy. Nevertheless, the recognition of the concept is already helping to widen the space for local and indigenous peoples and communities to manoeuvre and project their voices, represent their interests and fight for their rights. Indeed, the increasing recognition of indigenous and other community-based rights and the devolution of some administrative responsibilities for public forestlands to communities are two of the most important trends in forest management around the world.

A study by Forest Trends of 24 countries representing 93 per cent of the world's remaining natural forests indicates that there are at least 246 million hectares of forests officially owned by indigenous and other communities and at least 131 million hectares of public forest officially administered by indigenous and other communities in developing countries. In total, this is equivalent to about 22 per cent of all forests in developing countries – three times as much forest as is owned by industry or individuals (White et al, 2002). This is expected to double by 2015 and to exceed the 250 million to 300 million hectares of forest currently in publicly owned protected areas. At a time when external donor assistance for forestry has been declining and shrinking budgets limit the ability of states to invest in forest conservation, communities have been investing significant amounts in forest conservation. Forest Trends estimates that community investment in developing countries is equivalent to or exceeds external assistance flows to the forest sector and public expenditure by government (Molnar et al, 2004). Communities are now demanding a more level playing field, a restructuring of fiscal incentives and disincentives, and more practical regulations that do not discriminate against community forests and small forest enterprises.

At the same time that increasing opportunities are opening up for communities, for many, decentralization has also exacerbated tensions and conflicts, largely with the state but also with other stakeholders, and often within communities themselves. The authors of this book have noted tensions between externally stimulated and self-initiated community efforts; between the right of eminent domain of the state and land claims of communities; between the diversity of community preferences and approaches to forest management and conservation and the imposed uniformity of the state; between entrenched bureaucracies versus assertive communities over balance of power; between customary tenure and the formal definition and allocation of rights in the legal system. Many of these tensions are illustrated in the country case studies, particularly in the three cases included in Part III – Mount Kitanglad, a protected area in the Philippines (Chapter 16); Chizvirizvi, a resettlement area adjoining a wildlife reserve in Zimbabwe (Chapter 15); and in the Western Highlands of Guatemala (Chapter 12).

The three cases also share other common features. As in all other cases in this volume, there is a gap between the rhetoric and the practice of decentralization. The nature and extent of power devolved to local communities have been largely defined and assigned by external actors, notably the local government and bureaucracy, and these have invariably tried to retain for themselves as much meaningful power as possible. All three cases argue for further decentralization of authority from districts or municipality to community-level organizations and citizens. In the case of Guatemala, and in other parts of Latin America, decentralization is seen as 'municipalization'. The dominance of elite groups with a vested interest in consolidating their influence and hold on power has largely prevented further decentralization to the village level. Sharing of revenues from forests is a particularly contentious issue, since in all cases little is retained or reinvested in communities who were the source. And in all three, communities were not given access to high-value resources, although in the case of Mount Kitanglad, the community is represented in the management board for the protected area.

Decentralization is only one of many simultaneous processes that interlink and affect communities. What may be interpreted as community responses to decentralization is often adaptation to this confluence of processes, some dating back to colonial history. Agrarian reform and agrarian conflicts tend to be a subtext to decentralization not only for local communities, but for governments as well. In all cases, but especially for Guatemala and Zimbabwe, community perceptions and responses to decentralization are enmeshed in their continuing struggle for access and tenure rights to forest and arable land. For the state governments, as well, there is a tendency to converge decentralization and land reform as a way of settling tensions and competition over land.

In communities' struggle for land tenure, recognition and rights to forests and wildlife, external actors play a critical role and can have considerable influence. The success of communities' efforts requires the support, cooperation and legitimacy provided by these actors, including those whose interests may at times be at odds with those of the community, such as the rural district council in Zimbabwe.

The ability of communities and their leaders to gain access to and interact with higher authority and other stakeholders, as well as build alliances, is critical. In the case of Zimbabwe, the community leader's personal contacts not only facilitated funding for their land-use planning initiative, but also apparently strengthened his legitimacy in the eyes of his community. When conflict arises among factions within a community, external allies and donors tend to play the role of arbiters, and through their support for one or the other, can legitimize one among competing factions. In the Philippine case, the discord between a relatively well-educated aspirant and a less-educated traditional leader and their supporters is ostensibly an issue about legitimate leadership of a traditional institution that has been revitalized and inserted into a new multi-stakeholder institution created to oversee the management of the protected area. But it is also an issue about how traditional communities and institutions ought to relate with newly created, deliberately designed institutions for multi-stakeholder interactions, often involving more powerful external actors. And it is an issue of how a community chooses to represent itself to the outside world and gain access to external resources, and what sort of skills community members and their leaders ought to have to effectively protect and promote their interests.

In some instances, as in the case of Zimbabwe, communities are able to shift the balance of power in their favour and demand greater accountability and greater service value from the local authority for their tax payment. In other cases, such as in Bolivia (Chapter 9), the ability of communities to make demands is linked to their success in acquiring political power through elections. In the Philippines, decentralization has helped people to obtain property rights to indigenous and ancestral territories and community forests. However, the processes for titling and formalization of these property rights have generally been bureaucratic and slow.

Cases in which communities successfully assert and claim their rights are still few and far between. With the proliferation of external forest stakeholders and the creation or extension into communities of myriad institutions related to the process of decentralization, communities who are not well prepared and organized risk being mobilized and co-opted only to serve external agendas at the expense of their own interests. In Chapter 17, Elías and Wittman describe such a case. In Guatemala's Western Highlands, the superimposition of rules and institutions over pre-existing local institutions for land and forest management has weakened these local institutions and eroded their control over their communal forest resources. In Zimbabwe, the superimposition of political party structures and the politicization of their community institutions have created conflicts and divisions that threaten their land-use management institution.

In all three cases, communities see the need for capacity-building and development of technical skills as critically important. The state and civil society must enable communities to effectively manage their resources as a necessary component of effective decentralization. Capacity needs to be developed at different levels, but especially at the local level. The process of building capacity has to be sensitive to local culture. And it has to allow for enough time and resources to enable communities to ably and confidently assume their new roles and responsibilities and to effectively represent their interests and advocate on their own behalf.

CONCLUSION

Over the past two decades, responding to a host of factors and driving forces, a significant number of countries undertook to decentralize governance of their forest sector. The manner and characteristics of their decentralization efforts have yielded a range of results and impacts, both positive and negative. Decentralization was initially thought to result in improved participation, accountability and overall democratization in forest governance, as well as to promote sustainable forest management. In reality, however, the conditions necessary for decentralization to deliver on these expectations have rarely, if ever, been provided. Decentralization is clearly not a panacea, nor is it necessarily efficient and equitable. Rather, it has often been accompanied by conflict, particularly during the initial stages as legal, administrative and other uncertainties exacerbate pre-existing tensions among competing stakeholders and interest groups.

Under certain conditions, however, decentralization can yield positive results, and – while not itself sufficient – can contribute to poverty alleviation and more sustainable forest use and help to foster conditions for improving governance, in general. One size does not fit all, and no single brand of decentralization works in all cases. However, relatively successful decentralization experiences share some common features. Successful decentralization has been linked to secure tenure, as well as secure fiscal, revenue and taxation powers; equitable access to forest resources; control over decision-making, commercial rights and market access; sensitivity to cultural traditions and local knowledge; and, where appropriate, recognition of ancestral rights for local communities (Report to UNFF, 2004). For decentralization to be effective, both bottom-up and top-down forces are critical. It requires clear enabling legal and policy frameworks, and basic institutional capacities, including the capacity for resolving conflicts and negotiating among different stakeholders with competing interests and unequal power. Effective decentralization also requires timely and widely available information, as well as resources and mechanisms for upward and downward accountability.

Decentralization requires balance, particularly with respect to the roles and responsibilities of the central and lower levels of government, and across sectors and agencies relevant for forest management. This balance, however, has to be dynamic; it needs to be renegotiated and adjusted periodically as conditions change. In some contexts, decentralization has been undertaken with a strong central governmental role, while in others, the central government has had a less active role. A crucial lesson from the Interlaken workshop has been the ongoing tension, no matter how well established a country's decentralization, among governmental levels in determining legitimate rights, responsibilities and resources. The important influence exerted by powerful stakeholders, including the forestry bureaucracy, needs to be recognized, and care needs to be exercised to avoid elite capture of benefits and authority.

Decentralization takes time and, thus, is better implemented gradually, allowing for institutions and stakeholder groups to learn and to adapt. It requires building consensus through an open, transparent and inclusive process; participatory decision-making; institutional, technical and human capacity-building; provision of adequate financial resources and incentives for investment; tailoring objectives to local contexts; and developing the flexibility to adapt to different situations and changing circumstances. National forest programmes can be important opportunities and mechanisms to develop and strengthen multi-stakeholder processes and to incorporate broad-based inputs into the planning, implementation and monitoring of decentralization. Priority must be accorded to empowerment and capacity-building of local communities to effectively manage their natural resources. Finally, decentralization costs money: adequate financial resources need to be allocated to support the process. Incentives should be appropriately structured to promote investment and reinvestment in the forest sector. Effective and beneficial decentralization should not simply transfer the burden of management, as so often happens now, but must provide net positive benefits to local communities.

Recognizing that 'decentralization takes place in very different contexts', workshop participants wrote in their report to the Fourth United Nations Forum on Forests (Report to UNFF4, 2004, p9) that it is possible to formulate principles or guidelines as a reference for its implementation, 'which must be adapted by each country based on its own national reality'. Participants then listed those principles:

- Establishment of a clear legal and policy framework, with a clear allocation of roles, responsibilities and resources, should occur, as well as clarity and consistency regarding strategy and implementation.
- Decentralization of powers and responsibilities to districts and municipalities should not proceed arbitrarily, but according to a clear set of rules and conditions.
- Decentralization requires accountability at all levels and corresponding multiple accountability mechanisms; elections alone are insufficient.
- Decentralization should recognize, work with and strengthen representative, democratic institutions at all levels.
- Decentralization in the forest sector should not be implemented in isolation from a general national forestry strategy, such as national forest programmes.
- Decentralized forest management should be based not only on controls, but also on incentives; rules that cannot be enforced should not be made.

- Decentralization and the implementation of national forest programmes should include monitoring and evaluation with clear, specific variables and indicators.
- Rights and responsibilities must be accompanied by adequate resources and capacity-building.
- Decentralization should be based on transparent horizontal and vertical information flows and dialogue, including across sectors.
- Decentralization should both benefit from as well as enhance social capital, increasing coordination and trust among different levels and sectors.
- Local peoples must have a voice, and decentralization should take into account livelihoods, ways of life and improving the economic well-being of these peoples, as well as address inequities such as those relating to gender. Efforts must be made to raise and include the voices of special groups such as women, youth and indigenous peoples.

We conclude with our own, somewhat more general, lessons for effective decentralization, based on the findings reported in this book:

- Decentralization occurs within a particular historical, cultural, economic and political context, a context that uniquely shapes the process.
- Governance systems are characterized by an oscillation or balancing act between centralizing and decentralizing tendencies; it is a never-ending tension, and the balance differs in different contexts.
- Decentralization takes time, sometimes a long time, and requires attention to capacity-building, people's participation in decision-making and flexibility to adapt to changing circumstances and different situations.
- The decentralization process works best when there are reinforcing societal pressures from both the 'top' and the 'bottom'.
- Effective decentralization requires both upward and downward accountability. Serious efforts are needed to avoid elite capture of benefits and rights.
- Successful decentralization has been linked to secure tenure and access to forest resources, financial means and authority at lower levels, commercial rights and market access, and sensitivity to cultural traditions and local knowledge.
- Decentralization in the forestry sector is intimately linked to such processes in other sectors.

We hope that these suggestions are helpful in striking the right balance between decentralization and centralization, and that the great potential that exists for strengthened democratization, greater equity and more effective governance will be increasingly met in the years to come.

References

Molner, A., Scherr, S. J. (2004) *Who Conserves the World's Forests? Community-driven Strategies to Protect Forests and Respect Rights*. Forest Trends and Ecoagriculture Partners, Washington, DC

Report to United Nations Forum (2004) Decentralization, Federal Systems in Forestry and National Forest Programs: Interlaken Workshop. Report to United Nations Forum on Forests 4 (UNFF4), 27–30 April, Interlaken, Switzerland

Sithole, B. (2004) 'Becoming men in our dresses! Women's involvement in a joint forestry management project in Zimbabwe', in Colfer, C. J. P. (ed) *The Equitable Forest: Diversity, Community and Resource Management*. Resources for the Future and Center for International Forestry Research, Washington, DC

White, A. and Martin, A. (2002) *Who Owns the World's Forests? Forest Tenure and Public Forests in Transition*. Forest Trends Centre for International Environmental Law, Washington, DC

Wollenberg, E., Moeliono, M., Rhee, S., Limberg, G., Iwan, R. and Sudana, M. (2005) 'Between state and society: Local governance, conflict and forests in Malinau, Indonesia'. Paper presented at the Workshop on Community-based Forestry in the Next Decade: Towards an 'Agrarian Approach?', Humboldt University, Berlin, 8–9 October 2004, submitted to *Journal of Forest Policy and Economics*, forthcoming

Index